Beyond the Mountains

Beyond the Mountains

COMMODIFYING APPALACHIAN
ENVIRONMENTS

Drew A. Swanson

The University of Georgia Press

Athens

A version of chapter 3 appeared as "From Georgia to California and Back: The Rise, Fall, and Rebirth of Southern Gold Mining" in *Georgia Historical Quarterly* 100, no. 2 (2016): 160–86. Reprinted courtesy of Georgia Historical Society. A version of chapter 6 appeared as "Marketing a Mountain: Changing Views of Environment and Landscape on Grandfather Mountain, North Carolina" in *Appalachian Journal* 36, no. 1/2 (2008): 30–53. Reprinted courtesy of *Appalachian Journal*. A version of the epilogue appeared as "Mountain Meeting Grounds: History at an Intersection of Species," in *The Historical Animal*, edited by Susan Nance (Syracuse, N.Y.: Syracuse University Press, 2015), 240–58. Reprinted courtesy of Syracuse University Press.

© 2018 by the University of Georgia Press
Athens, Georgia 30602
www.ugapress.org
All rights reserved
Set in 10.5/13.5 Adobe Garamond Pro Regular by
Graphic Composition, Inc., Bogart, Georgia

Most University of Georgia Press titles are
available from popular e-book vendors.

Printed digitally

Library of Congress Cataloging-in-Publication Data

Names: Swanson, Drew A., 1979– author.
Title: Beyond the mountains : commodifying Appalachian environments / Drew A. Swanson.
Other titles: Environmental history and the American South.
Description: Athens : The University of Georgia Press, [2018] | Series: Environmental history and the American South | Includes bibliographical references and index.
Identifiers: LCCN 2018019190| ISBN 9780820344874 (hardback : alk. paper) | ISBN 9780820353975 (ebook) | ISBN 9780820353968 (pbk. : alk. paper)
Subjects: LCSH: Human ecology—Appalachian Region, Southern—History. | Natural resources—Appalachian Region, Southern—History. | Appalachian Region, Southern—Environmental conditions.
Classification: LCC GF504.A5 S93 2018 | DDC 333.70975—dc23
LC record available at https://lccn.loc.gov/2018019190

To Cindy, Mom, and Dad,
three "mountain" Swansons

CONTENTS

FOREWORD

Regional histories are hard to write. To do it well, authors have to link together two impulses that may seem contradictory on the surface: what is unique and what is common, what elemental quality of the place lies apart from the world around it and what is a product of it. Readers generally come to histories of an area, whether the Mississippi Delta or the Mekong Delta, with some boiled-down sense of what has defined that place, and it is the historian's challenge to explain what aspects of those regional preconceptions come from a place's insularity and what aspects are derived from its connections to the people and places around it. Regional environmental histories are even more complicated. Scholars must explain the ways in which the natural world shaped, even defined, a place, while stopping well short of explaining the sum of human action solely as the result of natural forces. Unfortunately, regional determinism and its environmental cousin have been powerful forces in popular and even scholarly conceptions of places, conceptions that have been particularly trenchant in the American South.

This brings us to the Appalachian mountain range, a locale to which Americans have ascribed all kinds of ideas about both human culture and nature. While academic scholars have contributed book after book pointing out the nuances of the region's society and economy, the popular image of the people and places in the mountains remains relatively flat. The rugged and harsh environment of the southern range, the old stereotype goes, produces people and culture that reflect those forces simply. In other words, nature determined Appalachian culture and Appalachians are constrained by that nature. The region lives in the American mind set apart from the rest of the nation.

What follows is a history of Appalachia that confronts regional and natural determinism head-on. Drew Swanson seeks to understand the uniqueness of the southern Appalachians through its many connections to the world around it, especially by exploring those people and institutions that came into the region in search of natural assets that could be taken away, packaged, and sold. At the center of this story—really a series of stories—is this commodification of the region and its plants, animals, minerals, people, weather, and views. For Swanson, to understand Appalachia means to appreciate not only the diversity of environments contained within it, but also the breadth of ways that people have shaped and been shaped by these forces. The various ways people interacted with nature and with the commodities that they extracted from it teach us, then, not just the varied *nature* of the southern Appalachian range, but also how the region was a part of the historical forces at work in the United States and across the globe.

The book's importance to southern environmental history is revealed in both the details of Swanson's telling and the sum of its parts. Though the chapters are chronological, each is centered on a commodity. Therefore, as he moves through time, the reader sees how a single aspect of nature—whether deer skins, gold, electricity, even tourism—is both created by its connection to the outside world and in turn influenced by outsiders' understanding of it. The approach works in great measure because of his insistence on the range of people, places, and things that should be considered commodities. Coal and salt, he argues, are not as different from electricity and tourism as we may have believed. His characters are not only the crags and valleys or the flora and fauna but actual humans, almost all of whom were trying to get something from those mountains. There are lots of scientists in Swanson's mountains— botanists, geologists, biologists, pedologists, herpetologists, and nuclear physicists—but their work appears alongside the more traditional historical characters we associate with Appalachia—the farmers, miners, and soldiers. The result is an argument that is not as obviously environmental as one might expect. Likewise, Swanson's refusal to concentrate only on those commodities that are already well understood, coal and timber for instance, makes the argument all the more important. Indeed, what I think is so refreshing about this book is the way that Swanson pitches the history of the Appalachian environment as non-determinist. This offers a break from what most environmental historians and Appalachian scholars have said about the region.

Having grown up on the edge of the Appalachian range, Swanson begins and ends the book with his own tromps through the woods and hikes in the mountains. His understandings of the landscape undergird much of the book, from vivid descriptions of vistas to the thrills of uncovering an endangered salamander under a rock. Indeed, Swanson has emerged as one of the most productive and influential voices of southern environmental history in large part because of this ability to blend landscape and environmental history with more traditional social and economic narratives. This, his third book and his second in the Environmental History and the American South series, rises to the challenge that few scholars have taken on, that of a complex environmental history of the southern Appalachian range. It finds good company in the series.

Alongside histories of plantations, oystering, soil erosion, and many other topics, *Beyond the Mountains* joins an impressive set of books that take similarly long views of human efforts to shape and use the natural worlds within the American South.

James C. Giesen
Series Editor, Environmental History and the American South

ACKNOWLEDGMENTS

In both undergraduate and graduate studies I was blessed with a set of teachers in Appalachian history and nature who stimulated my love for the region, including John Inscoe, Jonathan Sarris, Gene Spears, Allen Speer, Stewart Skeate, and John Alexander Williams. This book would not exist without their models.

A number of scholars and friends have made this a stronger book and exhibited a good deal of patience in listening to me over the years. Prominent among them are Mark Hersey, John Inscoe, Noeleen McIlvenna, Sirisha Naidu, Steve Nash, Kathryn Newfont, Tom Okie, Jesse Pope, Bill Storey, Paul Sutter, and Bert Way. Wright State University has proven a welcoming home, full of supportive colleagues; Jonathan Winkler and John Sherman expressed particular interest in this project. University professional development funds also made possible travel for research and presentations. At Wright State I have twice taught a course on Appalachian history, and each group of students pushed me to tell different stories, refine my arguments, and think more broadly about the region. Commenters at a number of conferences and workshops—including the annual meeting of the Appalachian Studies Association (2014 and 2015), Ohio University's Center for Contemporary History (2016), the Workshop for the History of Environment, Agriculture, Technology, and Science's annual meeting (2009), and Wright State University's History Department Faculty Forum (2017)—asked probing questions and helped me avoid some embarrassing blunders. I am also grateful for the journals and press that granted permission to republish small portions of this work: *Appalachian Journal*, *Georgia Historical Quarterly*, and Syracuse University Press. Editors Sandra Ballard, Glenn McNair, and Susan Nance assisted greatly in shaping those iterations of the work.

I have been helped along the way by an assortment of excellent archivists and records managers. I found assistance and friendly service at Appalachian State University's W. L. Eury Appalachian Collection; the University of Georgia's Hargrett Rare Books and Special Collections Library; the branch of the National Archives at Morrow, Georgia; the Georgia State Archives; Grandfather Mountain Park; and Lees-McRae College Library's Appalachian Collection. Librarians at Millsaps College and Wright State University proved invaluable as well, tracking down vital materials through interlibrary loan. I am also especially grateful to the University of Georgia's Special Collections Libraries and Katherine Stein for transforming a portion of my research on gold into a museum exhibit in 2017.

The University of Georgia Press has proven a wonderful publisher (again). Two anonymous reviewers offered insightful critiques of the draft, and series editor Jim Giesen asked probing and incisive questions that forced me to buckle down and carefully think about what it was that I was trying to say. Much that is good in what follows comes from his guidance. Mick Gusinde-Duffy did a wonderful job of steering the project through various editorial shoals, Melissa Bugbee Buchanan and Beth Snead facilitated production, and copy editor Chris Dodge beat a path through the thicket that is the English language.

As always, my largest debt and biggest thanks is owed to Margaret, Ethan, and Avery, who let me spend so much time traveling to the mountains, both in body and mind. You have my deepest love.

Beyond the Mountains

Introduction

A Constant Arcadia

TURKEYCOCK MOUNTAIN, part of the first wave of the Blue Ridge chain, rose up within sight of the backyard of my childhood home. As mountains go, it was not particularly impressive (less than two thousand feet in height, with a fairly gentle slope, and, lacking any definitive peak, it was more a ridge than a mountain), but it loomed large in my imagination. As I mowed the grass, milked the cows, or worked in the garden, it was never far from sight, a long blue-green forested mass lining the western horizon. I was fascinated by it and the other mountains beyond, rising wave upon wave, and I found myself always looking toward them rather than east into the rolling hills of the Virginia Piedmont. It was not the pull of mountain history—replete with tropes of hillbillies and moonshine and feuds and forgotten ancestors—that attracted me, however, though those stories interested me then and even more so now. Indeed, my ancestors had been a part of the region's real past, first moving to the edge of the Blue Ridge, according to family tradition, shortly before the American Revolution. Apparently many of them liked it enough to stay for more than two centuries. I was lured, rather, by Turkeycock's promise of nature. For a boy who lived in an outpost of the "flatlands," the mountains rising from the Piedmont seemed a wild place, territory to hunt and fish and roam, a land to escape modern life in favor of a constant arcadia.

To this end, one of my favorite pastimes when I was twelve or so was to go adventuring on the slopes of Turkeycock, a distance of several miles from home. I would cut across our pasture and woods and head out, crossing the back fields of a neighbor's farm, edging over the waters of Turkeycock Creek on a downed log or crossing on the shoulder of a county road bridge, and huffing and puffing my way up the northern shoulder of the mountain, squirrel hunting or bird-watching as I went. At the top of the ridge is a low cliff, twenty or so feet high, that looks out to the west. As I sat on the rocks to eat a snack and catch my breath, feet dangling over the edge, I would watch the clouds move over the Blue Ridge, and beyond them the Alleghenies, and imagine what it would be like to roam that far, to push into the great gray-green forest like some pioneering woodsman, Jim Bridger come back to life, living off my wits and the bounty of nature.

Of course, that land to the west was not truly wild, perhaps no more so than the rolling farmland to the east that I had tramped on my way to the foot of Turkeycock. Nor was it monolithic. The corner of the Blue Ridge I could see from the cliff was just as varied as other stretches of the South, a region whose variety has often been undersold. Appalachia as a cohesive place is just as much an organizational conceit as the "solid South." On the western side of Turkey-cock the valley between it and the next ridge, called Chestnut Mountain, con-tained the same sort of farms as surrounded my house: plots of tobacco, corn, wheat, and cattle, broken up by second-growth woodlots, farm ponds, tracts of cutover land, paved roads, and dirt lanes, with a small orchard here and there. If I could have seen farther, could have looked southwest beyond the New River Valley, I would have spied a similar mosaic in the valley of East Tennes-see and smaller pockets of diverse agricultural land throughout western North Carolina, southern West Virginia, eastern Kentucky, the Cumberland Plateau, upstate South Carolina, northern Georgia, and northeastern Alabama.

The woods themselves are equally varied. The forests surrounding Turkey-cock's cliffs are little different from the woodlands of the western Virginia Piedmont, composed as they are of hickories, white and chestnut oaks, various pines, and beeches, with a scattering of mountain laurel and huckleberries in the understory signifying the beginnings of the mountains. Traveling west for a few hours, however, reveals an increase in eastern hemlock, white pine, and sugar maples. Explore the high reaches of western North Carolina and you enter a completely different forest, a remnant of subalpine woods left by the retreat of the last ice age, where red spruce, Fraser fir, and yellow birch dominate forests similar to those of eastern Canada. Still other peaks from West Virginia to East Tennessee lack substantial forests and are instead termed "balds," mountain meadows covered with Catawba rhododendron, low alders, and varied herbaceous plants. Wet coves in the Great Smoky Mountains are virtual temperate rain forests. Among the most varied woodlands in North America, they boast stands of massive tulip trees, basswood, and great rhodo-dendron. Dry ridges just a few miles away often support scraggly Table Moun-tain pines and Allegheny sand myrtle, plants able to survive in the thin, porous soil and the ever-present wind that has twisted them into dwarfed forms.[1]

Not all of Appalachia's diverse terrain is so attractive or natural-appearing. In a swath of mountains rich with coal seams, extending from southern West Virginia through eastern Kentucky and southwestern Virginia and into Ten-nessee, mining companies have ripped apart the ridges themselves, employing the techniques of mountaintop removal, cutting the tops off peaks with explosives and heavy machinery to get to bands of coal and shunting everything else (labeled "overburden") into adjacent valleys. The region's major rivers and their tributaries sport numerous deep reservoirs, impounded by federal agencies or power companies in the twentieth century, fringed with docks and second homes and on sunny summer weekends humming with pontoon boats and jet skis. Peopled have graveled and paved over other linear strips of mountain environments, especially in the richest bottomlands, to provide the roads that move goods and people into and out of the mountains. Farmers' plows still transform forests into cropland annually, even as other farms revert to woods, while millions of regional residents conduct similar work on a smaller scale, tending yards, gardens, athletic fields, and golf courses carved from older landscapes. Yet nature continues to work in all these varied terrains.

And just try to predict the weather. Portions of Appalachia are decidedly southern, full of piney ridges that bask in humid summers and mild winters. Spring tornados and summer thunderstorms rise in the flatlands and hammer the flanks of the mountains, while the ragged remnants of tropical systems can push into the region from the east, south, or even west, bringing torrential rains and flooding. Other landscapes are more subalpine than subtropical, home to spires of rock and earth where fog freezes into rime ice on contact with the ground and winter snowstorms lead to an occasional whiteout, in which one might be forgiven for imagining the wilds of Alaska rather than the forests of North Carolina or West Virginia. Even on the slopes of a single mountain, the foot often lies in balmy sunshine, while wispy fog wraps the peak in a Scottish chill.

This diversity does not end with farms and forests, mines and highways, clouds and sun. In addition to its famed stretches of wild land, Appalachia is dotted with urban areas. The nearest substantial metropolitan area when I was a child—the place where we went to experience city life—was Roanoke, Virginia, roughly an hour's commute into the mountains. Many Appalachian people since World War II have lived in modern, postindustrial landscapes, places characterized by service industry jobs and tourism. They occupy tidy ranch homes, pseudo-log cabins, and condominiums in the suburbs and exurbs of cities like Knoxville, Chattanooga, Asheville, Roanoke, Charleston, the Tri-Cities (Johnson City, Kingsport, Bristol), and Huntington, as well as many other smaller towns. The people who live in these neighborhoods—and indeed throughout the region's rural areas—are also more diverse than stereotype suggests. They are black and white, descendants of slaves as well as slave owners. They are of Scots-Irish stock but also with recent ancestors from Hungary or Italy. They are Protestant, Catholic, or even agnostic, straight and gay,

Americans and Appalachians all. Appalachia was and remains many things, many places, many people, many natures.

∽

The object of this book is twofold. First, it engages with ongoing conversations about the degree of historic and current Appalachian isolation, supporting recent scholarship arguing that the southern mountains have been more connected to the broader world over the past few centuries and less exceptional than stereotypes long suggested. Indeed, on the cultural front this position has now become the accepted one, and the need to challenge the idea of Appalachian cultural exceptionalism may be waning.[2] There remains work to be done, however, in stripping regional scholarship of its tendency toward environmental determinism, in dispersing the "spell of the wilderness" that Berea College professor James Watt Raine claimed hung over southern Appalachia a century ago.[3] Regional studies that are remarkably sympathetic to Appalachian people's agency and diversity, indeed, their modernity, still have a tendency to treat Appalachian environments as overwhelmingly influential and somewhat monolithic. Second, and more originally, I argue that the region's connections often came from the use of (and thinking about) its nature, the very forces to which many observers attributed regional isolation. Although it is true that mountains could form physical barriers against the outside world, those same mountains' natural resources just as often encouraged economic, social, and cultural connections between people.

To get at this history of connectedness through a varied nature, the stories that follow trace the histories of a set of "commodities" in particular places in time. Historian Peter Coclanis provides a strict definition of a commodity as "a class or type of marketable good or service, members of which are generally sold in a rather undifferentiated, interchangeable manner and which at one end of the spectrum possess complete fungibility."[4] I use the term in a more basic, crude way, as a reference to something of value that is consumed in quantity. Some of the commodities in this book meet Coclanis's stricter definition—things like gold, salt, leather, tobacco, and coal—while others are more amorphous—transportation, scenery, power—yet no less influential for their vague and shifting qualities. All of these commodities were in some fashion products of mountain environments and served over time to connect Appalachian people and landscapes to more distant populations. In many ways they reveal a long history of thinking about Appalachian natural resources in similar ways. Wendell Berry has thus drawn a direct line from the fur trade to the strip-mining of coal, arguing that the regional "economy is still substantially that of the fur trade, still based in the same general kinds of commercial items: technology, weapons, ornaments, novelties, and drugs."[5]

Some readers will no doubt take umbrage at my selection of commodities. For example, no chapter is specifically devoted to timber or to iron (though both appear in multiple places). Additionally, the chapter titled "Coal" focuses on the evolution and side effects of one form of mining—mountain-top removal—rather than on the details of underground mining, the region's contentious labor history, or the experiences of miners and their families. My defense is twofold. First, there are many excellent histories covering these topics. For example, scholarly books on Appalachian coal mining could fill several bookshelves.[6] Second, exploring more varied and perhaps lesser-known topics illustrates the diversity of the region and the multiplicity of its environmental connections to other places. That lesser-known commodities followed patterns similar to those of timber and coal, and in some cases predated the systematic exploitation of those resources, suggests the power of these ways of thinking about and using nature in the southern mountains.

To emphasize this second point, I argue that Appalachian history is not defined by any particular environmental history. This is not to say that environments were not important; to the contrary, they were often extremely important. Kate Brown has recently admonished her fellow historians to pay more attention to material places, since "history occurs in place, not, as historians commonly believe, in time. . . . The fusing of spatial and temporal metaphors derive from the fact that time is the tracking of human action across space, which itself is a moving target."[7] Appalachia—the place—was as a whole largely rural and heavily reliant on agriculture, making rural environments in particular of great economic, ideological, and cultural importance. Within these environments there were tangible biological, chemical, geological, and meteorological forces that shaped life, and history that ignores them does so at the author's peril. Environmental history is useful in Appalachian history at least as much as it is in any other regional study. The crux for this region, as for so many other parts of the world, is that no single environment links all of the region's history. This assertion also makes irrelevant the ongoing, and, I believe, unproductive debates about how best to bound the region.[8]

Casting a shadow over these debates about regionalism and the environment are the mountains themselves. Do they furnish a physical, definitional structure to the region? The title of the sole synthetic environmental history of Appalachia, Donald Davis's *Where There Are Mountains*, suggests that the region's terrain proved an environmental unifier of sorts.[9] And the fact that the region's moniker has long relied on the name of its predominant mountain chain is perhaps even more convincing.[10] But one of environmental history's central assertions is that particular material environments matter a great deal, and on the ground Appalachia is a plethora of landscapes rather than a monolithic mountain chain. To distill regional history to the argument that "mountains shape cultures" is both reductionist and deterministic. In this sort of environmental explanation, what separates or differentiates the Unaka or the Unicoi Mountains of the American southeast from Italy's Alto Adige or the Peruvian Andes? To be sure, world-systems theory argues that Appalachia and many other mountainous regions across the globe are much alike, places made peripheral and subservient to core economic and cultural hubs at least in part because of the challenges of mountain environments.[11] There is a bit of truth in this notion, but it obscures as much as it reveals about the essence of place. It is, as historian John Alexander Williams notes, a framework of "relative inutility."[12] This book argues instead that on the ground environments deeply mattered but that local environments mattered more than broad, abstract categories, as was the case across the nation and world as a whole.

These collected stories—and the disparate environments they reveal—ultimately reinforce what Appalachian studies scholars have been telling us for years: Appalachia is not and was not a place outside of the American experience. Yes, the southern mountains were relatively remote, isolated, and under-developed, but the key term here is relatively.[13] What the following environmental histories convey is that what usually mattered most was the relationship between particular mountain environments, local ideas, and other places. Appalachian history is a story of specific relations more than isolation. People in the past rarely obsessed about an overarching mountain environment; instead they cared about specifics. As we will see, Cherokee people and European traders cared about the qualities of deerskins and trade paths, Union officers focused on geographical defenses and lines of retreat, tobacco farmers thought in terms of soil types and frost-free nights, and mine engineers dwelled on the best way to push the top of a specific mountain into the adjoining valley.

Despite the lack of a singular nature in Appalachian history, big ideas about regional environments—even if flawed—did matter. As William Cronon and others have argued, historians still need to take the fractious mass that is history and shape it into stories, because humans are habituated to understand the past (as well as the present) through the narrative form.[14] A chronicle that simply brought together a set of past environmental experiences from the southern mountains might in some respects reflect a certain historical reality, but it would offer little meaning. And so, despite this book's argument that Appalachia was many places, each connected to the outside world as much as to one another, it also accepts that the idea of Appalachia as a singular place mattered. If topography and actual relationships across space were crucial in Appalachian history, then mental geographies were equally important in forming the region's past. From scholars Henry Shapiro, Allen Batteau, and others, we know that Appalachia has been a terrain of the mind as much as physical ground.[15] From the first days of Euro- and African-American settlement, people carried particular ideas about nature with them into the mountains, although these ideas were fluid. In many cases they proved at odds with one another and with Native American ideas, and they changed over time. The various ways of seeing mountain environments appear throughout the following pages. French and English woodsmen saw a landscape of fur and empires' boundaries, botanizers wandered through a repository of rare plants, miners envisioned mountains of precious metal, Union and Confederate officers thought of salt as they planned operations, tourists sought healthy environs and dramatic views, loggers and colliers believed the measure of mountains grew from their soil or hid beneath it, farmers sought soils that grew light-colored tobacco, urban boosters touted a land suited to industrial growth, atomic scientists saw Appalachia as a safe place to hide their work, and mountaintop removal opponents protested that their homeland was more than abstract natural resources. In addition, national ideas about the region intersected local ones. Looking at Roanoke, Virginia, the subject of a chapter on urban Appalachia, we see post–Civil War national ideas about the worth of merging northern capital with southern natural resources encounter local ideas about the best site for rail lines. For all the potential conflict between these various views, they were nonetheless important in shaping physical Appalachia.

This book's focus on commodities brings together these two ways of conceptualizing Appalachian nature—as local specificities and broad regional resources—illustrating the contradictions inherent in Appalachia's environmental histories. Specific ecosystems shaped history on the ground, influencing where people lived and moved, how they worked, and how they understood themselves and the world around them. Commodities came from actual places, where topography, weather, and nonhuman life influenced their form, extraction, harvesting, and transportation, and people who produced these goods well understood these facts. But residents of Appalachia and elsewhere also came to view many of these resources as commodities, imagining them as part of national and international economies, tradable across distance and sometimes even fungible. In some ways these external perceptions of the Appalachian environment eventually came to bear more weight than internal ones, as the rising tide of capitalism valued trade and profit above tradition and community.

These economic ideas also came to influence the nation's imagination, as stereotypes about backward Appalachian people themselves became a sort of pop culture commodity. If residents of various valleys and ridges thought of their local environments as productive or poor, homey or inhospitable, outsiders increasingly created a homogenous stereotype of mountain land and people. In these portrayals the American idea of Appalachia as a place and people emerged. Thus, over time, real connections spawned visions of isolation. This was the land of abundant natural resources and people too isolated to make full use of them, a land of "our contemporary ancestors" where the hills and dales evoked Scotland and Ireland.[16] It was a place where natural descriptions and human stereotypes fused in terms like "hillbillies" and "mountaineers."[17]

Tourists and local color writers of the late nineteenth and early twentieth centuries like Mary Noailles Murfree and John Fox Jr. carried these portrayals to the reading public, and Hollywood seized upon this naturalized poverty and reproduced it for even broader audiences in the decades that followed. The results are now American pop culture icons: *Snuffy Smith* and *Li'l Abner* comics, *Beverly Hillbillies* and *The Dukes of Hazzard* filling evening TV schedules, and, in both novel and movie form, *Deliverance* portraying the deviance said to come from too great an isolation. Nor has this linkage of environment and culture faded with time. The contemporary entertainment industry continually rehashes these themes for eager audiences. In addition to film versions of *The Beverly Hillbillies* and *The Dukes of Hazzard*, examples include MTV's *Buckwild* series, a purported "reality" show that promises to reveal the debauchery of West Virginia teens, and *The Wild and Wonderful Whites of West Virginia*, a pseudo-documentary by Johnny Knoxville (producer of *Jackass*) that follows a similar script for a mountain family plagued by drug abuse and violence.[18] An even more extreme example is the *Wrong Turn* franchise of horror movies (at this writing on its sixth installment). Gorier and less artistic versions of *Deliverance* with its focus on the cultural regression that can accompany loss of contact with a broader world, the first film featured lost tourists in the rural West Virginia woods pursued by "cannibalistic mountain men grossly disfigured through generations of in-breeding."[19]

More serious and balanced works that nonetheless accept the deterministic power of Appalachian nature have recently found wide readership. Jeff Biggers's *United States of Appalachia* offers a thoroughgoing corrective to ideas of the region's cultural backwardness, but it often does so by implying that the challenges of nature in the southern mountains produced a vibrant, creative, and flexible people by forcing them to overcome environmental obstacles. In his account, Appalachian people were simply a different kind of exceptional rooted in their environments.[20] J. D. Vance's bestselling memoir, *Hillbilly Elegy*, offers little such hope in the power of exceptionalism, instead introducing readers to a family history of poverty and pessimism that Vance argues is rooted in Appalachian history. In white, Scots-Irish Appalachia, he declares, "If ethnicity is one side of the coin, then geography is the other," and he concludes that "hillbillies" are "more socially isolated than ever before" in the hills and hollows of the southern mountains. These stereotypes undergird the book, even as it sketches a family caught in the economic flows of modern America, moving back and forth between the mountains of eastern Kentucky and industrial Middletown, Ohio.[21]

Environmental histories intent on describing the regional environment and its effects, in search of mountains as synecdoche for a bigger Appalachia, do similar, if inadvertent, violence to the region's kaleidoscope of people, cultures, and natures. And my childhood visions looking west from Turkeycock Mountain were firmly rooted in these old yet ongoing stories of Appalachian nature and history, to the extent that I then understood them. They combined a set of ideas and images with real landscapes, in ways that blurred the past and present. They transformed local environments and specific histories into a region and a collective past, for better or worse. My constant arcadia was a matter of perspective, a product of the relationship between where I lived—my Piedmont gaze, if you will—and the view to the west. What might I have actually seen had my feet followed my gaze, had I been able to move deeper into the mountains and into history, past the broad strokes of "Appalachia" and into its particular stories? What follows is an effort to tease out a few of those tales.

CHAPTER 1

Leather

The Deerskin Trade in the
Southern Mountains

IN THE CLOSING DAYS OF 1761, Lieutenant Henry Timberlake noted in his journal that he had seen the ruins of Fort Loudon. Timberlake was in the mountains of what would become East Tennessee as part of a British delegation certifying that peace had been brokered with the Cherokee people, who had rebelled against their alliance with the British Empire in the French and Indian War. In addition to the fallen fort's serving as a landmark, Timberlake viewed it as a reminder of the stakes of his party's endeavor. The British had built Loudon five years earlier to guard the frontier from the French and their Shawnee allies, and a Cherokee war party had taken the fort only months earlier, killing part of its garrison. By the time Timberlake gazed upon the ruins of the fort, the relationship between the Cherokee people and the southern colonies had long been politically important to both parties, and just as significant was its commercial import. The colonial economies of South Carolina and Virginia, and to a lesser extent Georgia and North Carolina, had for decades relied on profits produced from the export of deerskins—many taken by Cherokee hunters—and peace offered the hope of renewing that trade. Timberlake believed that he and his fellow soldiers were ensuring a refreshed world of commerce rather than witnessing the end of an era.[1]

The Anglo-Cherokee War, as a part of the broader French and Indian War (itself one theater of the intercontinental conflict known as the Seven Years' War), is well recorded, but its intimate connections with the century-old deerskin trade are less often highlighted. Historians frequently treat the commerce in hides in the southern mountains as a transient frontier enterprise, part of a hardscrabble economy only pursued before the arrival of more permanent farming, timbering, or mineral extraction. This interpretation does a disservice to the significance of the southern fur trade, which was an economic activity central to Indian life for nearly a century and did much to drive colonial understandings of the mountain frontier and eventually encourage Euro-American settlement in the region. Indeed, the exploitation of Appalachian

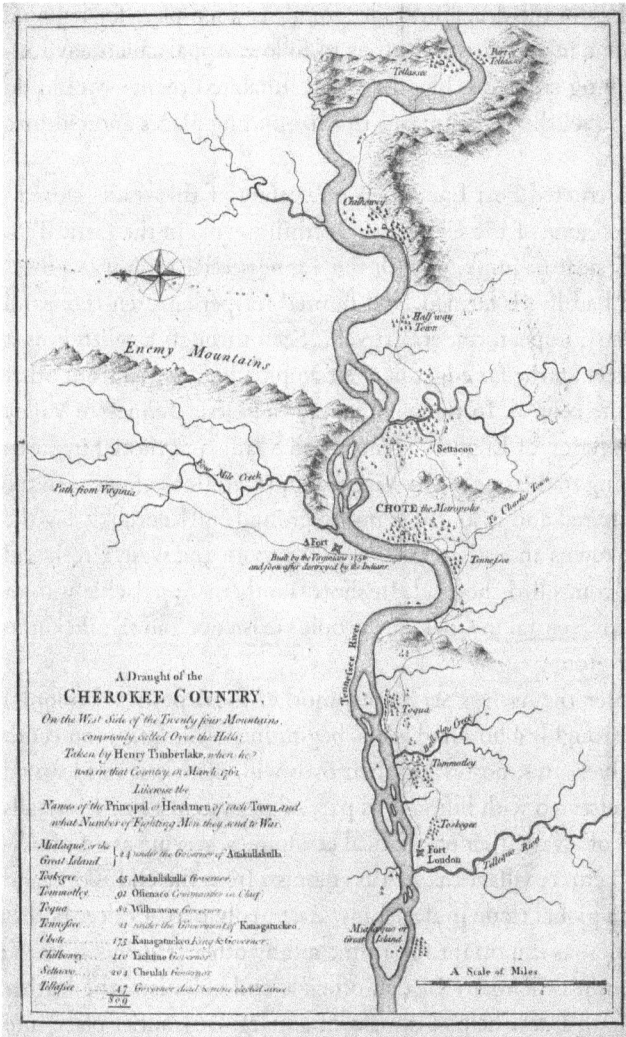

FIGURE 1.1. Henry Timberlake sketched a map of his journey through Cherokee Territory at the conclusion of the Anglo-Cherokee War. Note Fort Loudon in the lower right-hand corner. Henry Timberlake, *The Memoirs of Lieut. Henry Timberlake* (London: J. Ridley, W. Nicoll, and C. Henderson, 1765).

resources undertaken in the deerskin trade would set a template for regional commercial exchange in the three centuries to follow. Appalachian environments—and the living creatures, like deer, that inhabited them—would do more to connect the southern mountains to surrounding places and cultures than to isolate them.

The British constructed Fort Loudon at one center of this trade, close to the village of Tellico (one of the Cherokee Overhill towns) in the Little Tennessee River Valley near its junction with the Tennessee River. It was a location where people had lived, farmed, and hunted for perhaps ten thousand years when the first European traders arrived.[2] Seen through English eyes it was a marginal place, on the far edges of their empire, but viewed from other vantages it was quite central. In the colonial era the Little Tennessee Valley existed at the intersection of English, French, and Spanish territorial interests and in an overlapping trade zone of the Virginia and Carolina colonies. To the north, across the shared and disputed hunting grounds of Kentucky, lay the powerful Shawnee towns and other peoples; to the south and west, Creek and Chickasaw people controlled the land. In short, southern Appalachia was an important crossroads, central to Cherokee people's existence and significant to a great many other groups.

White-tailed deer themselves served as another focal point of colonial North America for roughly a hundred years, beginning in the late seventeenth century. As living creatures, possessing their own will, instincts, and learned behaviors, deer intersected with humans as part of ecosystems. But whitetails also became a part of a varied set of political economies, serving as an important resource for Cherokee villages as well as colonists from the Old World and ultimately fueling a global trade in deerskins. Eventually parts of deer bodies became commodities, as important for a time as any other trade good in the southeast. In the minds of many British officials and colonists, the leather made from white-tailed deer helped transform a marginal frontier place into an important region, a crossroads of commerce that would influence people and empires.

Historians have treated this deerskin trade in various ways. Older histories often approached this commerce through the lens of adventurous and independent fur traders, by consequence depicting it as a rustic, frontier activity, destined for displacement by more civilized endeavors. In short, it was part of a romantic but transient frontier, the sort of phenomenon that fit neatly into Frederick Jackson Turner's theory of progressive American development. More recent histories have presented more nuanced takes, emphasizing the substantial economic importance of the trade for British, French, and Spanish colonies; exploring its devastating consequences for Native Americans, through the spread of disease, disruption of historic patterns of life, and alcoholism; and noting the intimate connections between the hide trade and warfare.[3] A few scholars, Wilma Dunaway most notably, have labeled the deerskin trade a crucial component in the establishment and spread of a capitalist ethos in North America. Dunaway views the "southern deerskin trade as the key mechanism by which Native Americans and their lands were incorporated into the world-economy" and sees the pursuit of hides as "the chief instrument of economic expansion during the early Colonial period." She likens the results to another version of a contemporary global phenomenon, that of the "putting-out system that destroyed traditional economic activities, generated dependency upon European trade goods, and stimulated debt peonage."[4] If equating the southeastern trade in whitetail hides to contemporary textile work in England is a bit specious, commerce in deerskins undeniably brought social, economic, and cultural revolutions to the southern mountains. The lives of Cherokee families, British colonists, European consumers, and millions of deer would be forever altered.

໔໑

At the moment when Europeans first set foot in the New World, an incredible number of white-tailed deer inhabited much of North America. They had adapted to a wide range of habitats, from the tropical forests of Central America to the deciduous woodlands of Appalachia to the frigid coniferous forests of what would become eastern Canada. Only the largely treeless Great Plains and drier portions of the American West were entirely devoid of whitetails. Descriptions of their abundance are numerous and often include a note of wonder, as one from 1682, when Carolinian Thomas Ashe wrote, "There is such infinite Herds that the whole country seems but one continued [deer] park."[5] Attaching a precise figure to anecdotal accounts such as this is impossible, but historians and wildlife scientists have nonetheless attempted to calculate the contact period whitetail population. In the most exhaustive study to date, Thomas and Richard McCabe estimated the number of deer at "24 to 33 million," and most other careful estimates fall within or close to their figures.[6]

These population numbers, if correct, reflect annual average populations more than a static continental herd of deer. Local populations then, as now, fluctuated widely, dependent on hunting pressure but also on a range of other environmental factors. For example, exceptionally hard winters, sustained drought, or lightning-sparked wildfires could cause regional or local populations to decline, while widespread tornado or hurricane damage, relatively common in the Southeast, could create additional edge habitats in mature forests, which could benefit local whitetail populations for a time.[7] Deer in turn functioned as a sort of "disturbance" in many eastern forests, like those of the Appalachian Mountains. Dense deer populations over time created particular sorts of forests. As one ecologist has written, "White-tailed deer may therefore be considered a 'keystone species' in deciduous forests, because they have a disproportionately large impact on the distribution and abundance of other species in forest communities and the resulting community structure." For example, whitetails favor white oak seedlings as browse, and their feeding suppresses that species, in turn "releasing" other trees, less-preferred ones such as sugar maples, to increase their stands. Over the span of decades, plentiful deer had the power to shape forests just as much as a hurricane or wildfire.[8] Cumulatively, these varied processes led to shrinking and swelling deer populations, always fluctuating, never in balance.

White-tailed deer were among the most important natural resources for southeastern Native Americans. For many, including the Cherokee people, deer served as the primary animal food source, but whitetails provided more than just venison. Deer hides offered several advantages over the skins of other mammals: they were strong enough to stand up to hard wear yet supple enough to make comfortable clothing, and deerskin remained pliable when wet, unlike buffalo hide. As a consequence of these qualities and deer abundance, hides were used for a wide range of clothing and other goods. They became moccasins and gloves, arrow quivers, cordage, tobacco pouches, rugs, drumheads, and even the balls used in Cherokee games. In addition, whitetail bones, antlers, sinews, and hair had many uses, from tool handles to bow strings to insulation. Even the brains of the animal were used, to tan its own hide.[9] These varied uses meant that Native Americans who relied on the species consumed a large number of whitetails each year, perhaps as many as nine or ten adult deer per person just for domestic use.[10]

Cherokee hunters took these deer in a number of ways. They pursued whitetails with bows and arrows, through the traditional means of solitary stalking or "still hunting" (waiting in ambush) (see fig. 1.2). But they, like other Native Americans throughout the whitetail's range, also used communal methods of hunting, often involving fire. Intentionally set fires forced deer to move, often driving them toward a waiting group of hunters. These regular hunting fires in turn reshaped mountain environments, in many ways making them more suited to white-tailed deer. Regular burning of the woods suppressed the woody understory, created additional edge habitat, and fostered the growth of many herbaceous plant species favored by deer. Together Cherokee hunters, deer, and the frequent fires used in hunting formed distinctive woodlands throughout the southern mountains.[11]

FIGURE 1.2. An imaginative European depiction of southeastern Indians stalking deer. Jacques Le Moyne de Morgues, 1603. Courtesy of the Library of Congress.

Their reliance on whitetails meant that Cherokee people held a belief that deer were tied to more than just the state of Appalachian forests; they were a part of the very fabric of culture, society, and the spiritual world as well. Their mythology emphasized the importance of deer, weaving them into creation stories and explanations for the origins of illnesses. They told tales of how all deer once existed underground, to be released one by one from under a rock in times of hunger, but mischievous boys set them all free, populating the earth with game but necessitating hunting for meat. And Cherokee stories recounted how careful prayer to a deer's spirit was necessary after a successful hunt, or else the spirit known as Little Deer would smite the hunter with chronic rheumatism, a crippling condition in a culture dependent on physical labor.[12]

Even as Cherokee hunting was drawn into networks of transatlantic trade during the eighteenth century, it remained freighted with a great deal of cultural and religious significance. Despite a growing dependence on the deerskin trade, there were cultural proscriptions against taking too much game and against extending hunting grounds too far, though these were not inviolable. Hunting trips could last as long as half a year, and hunters did not lightly undertake them. Preparation involved ritual abstinence and religious sacrifices, and symbols of the deer were everywhere: "seven deer skins folded" were part of the start of a hunt ceremony, as were divinations that might reveal "a great multitude of deers hornes" if the hunt was to be a successful one. During the hunt itself, deer tongues and spleens were sometimes offered as sacrifice, and hunters sat on deerskins in camp. Participation in the deerskin trade changed many things in Cherokee society—bringing economic intensification, pulling more members of society into the production of hides, and perhaps forming newly gendered divisions of labor—but it did not entirely eliminate the white-tail's place in the older Cherokee cosmology.[13]

As important as deer were to Cherokee life, prior to the hide trade with Euro-Americans the animals were merely one part of an economy that relied on both wild and domesticated resources. Farming lay at the heart of Cherokee life, which revolved around seasonal work in fields of corn, beans, and squashes. Colonial English visitors to Cherokee towns in the late 1600s and early 1700s remarked on the fertility of mountain valley soils at sites like Tellico, which had "the richest soil, equal to manure itself, impossible in appearance ever to wear out."[14] Valley town sites typically lay within close reach of the diverse forest ecosystems of the mountain slopes, in a region that contained "all major forest complexes," thanks primarily to varied exposures and elevation differences. These woods were rich in hickory nuts, acorns, chestnuts, black walnuts, and butternuts, as well as game, herbs, and edible roots, and the rivers and streams were home to a range of fish species.[15] The Cherokee people's ability to transform these natural resources into a relatively comfortable lifestyle impressed outside observers. One colonial traveler commented that a Cherokee hunter in "the extensive woods" could in short order "collect fire . . . make a bark hut, earthen vessels, and a bow and arrows; then kill wild game, fish, fresh water tortoises, gather plentiful variety of vegetables, and live in affluence."[16] As a consequence of their skill in using this varied and rich environment, populations at favored sites could be quite dense. Along the tributaries of the Little Tennessee River near Tellico, for example, archaeologists have discovered roughly two dozen sites of historic significance, the culmination of regional human habitation dating back perhaps as many as ten millennia.[17]

The Cherokee towns were relatively late entrants to the fur trade, at least on a large scale. They had sold some skins to Spanish traders during the first decades of the seventeenth century, but it was only when the English entered the fur trade later in the 1600s that significant numbers of deerskins left the southern mountains for distant markets. Virginia was the first English colony to pursue trade with the mountains. Its government sent two traders, James Needham and Gabriel Arthur, on an expedition to establish relations with the people of the southern Appalachians and secure a new source of cheap furs.

In 1673 they set out from Fort Henry, located on the falls of the Appomattox River (at modern-day Petersburg), in an effort to bypass Native American intermediaries, and reached the mountains where they conducted trade. Despite this early contact, prior to the 1710s almost all the fur that came from Cherokee territory still flowed through intermediaries like the Catawba and Occaneechi peoples who lived in the southern Piedmont.[18] As late as 1713, "the Cherokee were still 'little known' to the English," and this may have been due to an intentional decision on their part. Located between the colonial outposts of England, France, and Spain, the Cherokee towns had some contact with each European power, and they had been buffeted by the forces of disease, war, and displacement that accompanied the Columbian Exchange. There is evidence that Cherokee leaders sought to avoid full entanglement in the fur trade and the further problems that might come with it. Aside from vestiges of firearms and ammunition, there is little archaeological evidence of significant incorporation of European trade goods into Cherokee life before the 1710s.[19]

By the early eighteenth century, the Carolina colony became a strong competitor for the mountain fur trade, soon all but displacing the Virginians. Carolina's merchants sought a commodity that would make their colonial endeavor pay. Indigo and then rice would soon drive the colony's agriculture, but deerskins from the Piedmont and later the mountains as well as the Gulf Coastal Plain were a crucial ingredient in Carolina's early economic success. Charleston was well positioned to capture the bulk of this commerce from Virginia traders, as it was closer to the southern mountains, and especially to the lower Piedmont and the Gulf Plain than was the Chesapeake. In the early Carolina trade, the most important suppliers of hides were Creek and Chickasaw hunters, but Cherokee communities also increasingly became involved in the commerce.[20]

Early commercial journeys into the mountains were largely private endeavors, but South Carolina's colonial government soon moved to systemize and control the trade. In 1707 the colony passed its first rules governing commerce in hides, and in 1716, in the aftermath of the Yamasee War that had temporarily disrupted the fur trade, it declared a government monopoly on trading in deerskins. Although its restrictions were clear, this legislation never truly did away with private trading, and the monopoly remained in place for only three years before officials gave it up as unenforceable.[21] Despite this struggle over regulation, South Carolina's aggressive private and public trading quickly resulted in its control of the southeastern trade. Between 1698 and 1715, Virginia ports shipped an average of fourteen thousand hides each year, but Charleston exports dwarfed those figures. Over the same period roughly fifty-four thousand skins left South Carolina for England annually, and by the 1720s that number had swelled to more than a quarter of a million deer hides each year, with these official figures almost certainly undercounting the actual number.[22]

Export statistics as substantial as these ensured colonial efforts to maintain the valuable trade. In addition to the short-lived effort to establish a monopoly, the South Carolina legislature appointed commissioners to oversee relations with Native Americans and sent several expeditions to Cherokee territory to observe the fur trade and make recommendations on how it might be perpetuated and improved. A concern was a renewed effort by Virginia—which established a government venture named the Virginia Indian Company in 1718—to gain a larger share of the trade with Cherokee towns.[23] One important expedition in 1715, led by George Chicken (who would become a South Carolina commissioner of Indian affairs), secured Cherokee people's support for the English in the Yamasee War and strengthened trading relations. Chicken visited the Cherokee towns in an official capacity again in 1725, on a tour to inspect the honesty of colonial traders and the efficiency of the trade.[24] In 1730 Alexander Cuming renegotiated the trade and political alliances between South Carolina and the Cherokee government, promising a steady flow of guns and ammunition in exchange for skins and a rejection of French overtures.[25]

These endeavors managed to maintain a flow of hundreds of thousands of hides from the Southeast to colonial centers and European markets, and people along every link of the commercial chain put deer to a varied range of uses. English settlers on the frontier's edge used their share of deer, often in ways quite similar to Cherokee people's domestic consumption. They ate venison, even if they found it far too lean to be truly enjoyable, and they manufactured hides into "breeches, jackets, waistcoats or entire suits, and moccasins," as well used them for "stocking hats and gloves," parts of snowshoes, window coverings, whips, and to splice wire. Antlers became knife and tool handles, while hair served for bed stuffing and insulation.[26] The most important market by 1700, however, was in Europe, where the durable yet soft skins found a ready demand in the leather trade. North American whitetails became parts of military uniforms and English book covers (in some cases returning across the Atlantic as coverings of both soldiers and laws that endeavored to rule the whitetails' homelands). Artisans cut and stitched deerskins into jackets, shoes, saddles, work aprons, and gloves, and they wrapped them around trunks and boxes. English demand in particular was quite high in the eighteenth century, since a series of epizootic diseases had diminished European livestock herds, and to arrest the contagion England restricted trade in cattle and leather with the continent.[27] As a consequence, the best-quality American deer hides went to England, the next best often ended up on the German market, and inferior skins usually remained in the colonies.[28] It is also important to note that other animals were a part of the southern fur trade. Appalachian beaver—though not providing pelts as high in quality as ones from Canada or the Upper Midwest—became European hats, and Virginia exporters recorded appreciable sales of black bear, bobcat, fox, fisher, mink, muskrat, otter, raccoon, wolf, and even groundhog hides. Still, deerskins dominated the commerce.[29] From Williamsburg to London to Munich, goods made from Appalachian deerskins became a part of people's everyday lives.

In a reverse commodity flow, by the early eighteenth century European technology in the form of firearms and ammunition became a central part of Cherokee hunting, where it served to satisfy both Native American and European consumer desires. Guns only slowly replaced bows and arrows in the late-seventeenth-century Southeast, but new flintlock firearms revolutionized Indian hunting during the first decades of the eighteenth century. The modern guns were lighter, easier to fire, more water-resistant, and more accurate, and they quickly became the preferred tools of Cherokee hunters. Cherokee people traded for firearms and ammunition more than any other goods, and treaties like the one with South Carolina that promised two hundred long guns and ammunition in exchange for an alliance during the Yamasee War (1715) brought more firearms into circulation.[30] Just a few new guns could transform the efficiency of deer hunting in the Southeast. As one Choctaw hunter remembered, firearms radically changed his village's hunting practices: "In place of the large companies and laborious running, surrounding, and driving, men would sneak out alone and could accomplish more than twenty men could with the bow and arrow, and never go out of a walk[ing distance]."[31]

Flintlock guns made Cherokee hunters more efficient, but they also more thoroughly enmeshed their communities in distant networks of trade and credit than ever before. The only things hunters could purchase firearms with were deerskins, and increased hunting to obtain those skins demanded yet more firearms and ammunition. Likewise, efforts to extend hunting ranges led to increased conflict with other Native Americans, such as Creek and Shawnee peoples, who were also expanding their participation in the fur trade. In turn, these struggles encouraged Cherokee towns to purchase more guns for defense, a side effect of which was increased hunting pressure on deer. Each purchase and use of a gun ultimately advanced the logic and necessity of killing more deer and buying additional weapons. Colonial traders, especially South Carolinians, recognized this growing dependence on guns and periodically threatened to stop the supply of firearms or ammunition to the Cherokee towns as a powerful political tool. These cycles of debt and payment, hunting and war, grew stronger as the eighteenth century progressed. Eventually the older ways of hunting all but disappeared from Cherokee cultural memory, as older hunters passed away and left a generation accustomed only to taking whitetails with firearms.[32]

Another introduction, one with as least as much impact as firearms, was alcohol; together with skins and guns, rum would forge a tripartite trade in the mountains. From the first years of the trade, English merchants faced a quandary: How could they sell more goods to a relatively small population of Native Americans in order to acquire ever more deerskins? Cherokee people in particular favored durable wares like guns and kettles, which rarely needed replacing. And Cherokee populations did not increase through the early and middle decades of the eighteenth century; in fact, a smallpox epidemic, likely introduced by South Carolina fur traders, swept the Cherokee towns in 1738, killing perhaps half of the total population. Durable goods and a stable or declining population seemed to promise quick market saturation and little ongoing demand (except for ammunition and gun parts). The answer traders seized upon was rum. The spirit proved popular, was quickly consumed, and, in the mercenary logic of commerce, its addictive nature ensured "a highly 'elastic' demand," properties that made it the perfect trade good from English traders' perspective. As a consequence, over time more rum flowed into the mountains and more deer hides flowed out.[33]

Both Virginia and South Carolina governments recognized the potential for corruption, violence, and political turmoil that accompanied the rum trade and sought to prohibit it. To be sure, their concerns were more economic and political than moral. Early in the eighteenth century the colonies passed laws against selling rum to Native Americans, but these proscriptions proved ineffective. According to a number of accounts, alcoholism became a serious problem quite quickly.[34] In addition to rum's destabilizing societal effects, the elasticity of the demand for it created serious stress on Indian economies. By mid-century a gallon of rum commanded an average price of four good deerskins (more than 10 percent of the value of a musket), and eventually in certain locales two hides would buy only a quart of watered-down brandy.[35] Cherokee hunters could increase their efforts to some degree, staying afield longer and ranging farther, but with high rum prices, an ever-expanding market for alcohol, and a limited number of hunters and deer, debt became a serious problem. Indeed, the balance of trade quickly spiraled out of control. To give one example, "the 1711 rum debt alone" owed to South Carolina traders by southeastern Native Americans "amounted to approximately 100,000 skins, nearly two years' work for the Indian trappers" at the time. Many towns became permanently indebted to colonial traders, ensuring cross-cultural tension and an increasing motivation to kill ever more whitetails or to break trade agreements and repudiate debts. The Yamasee people owed the greatest share of the rum debt held by South Carolinians on the eve of the Yamasee War, and that debt contributed to the conflict.[36]

Trading corruption extended beyond intentional efforts to foster addiction and debt. In his second tour of Cherokee territory, Commissioner Chicken asserted that private traders ought to be more tightly regulated, as some of them would "say or do any thing among the Indians for the Lucre of a few Skines." Again, commerce as much as morals drove his concerns. He was particularly troubled that so many poorly cured skins were entering the trade since without proper tanning the hides often spoiled before reaching their final destination. When he informed Cherokee leaders of his concerns, they responded that they well understood proper preservation of deer hides, but traders' greed undermined the process. They declared that "it was intirely the White Mens faults and that some of them followed the Indians into the Woods for their Skines and that they love them so well that they do not care if they take them raw or any other ways." This claim highlighted the value that Cherokee labor added to raw deerskins. A properly dressed hide was one that had been washed, scraped of loose tissue, washed again and scraped of hair, rubbed with the animal's brains, then smoked over a low fire. It was a process that took more than a week but produced a durable, pliant skin of medium brown hue, ready for shipment or manufacturing into a wide range of goods. Bypassing any of these steps made for a hide of much lower quality. Chicken's response to these greedy shortcuts was to issue orders to all South Carolina traders forbidding the purchase of undressed hides. That he was forced to issue the decree multiple times indicates a lack of compliance by traders.[37]

For their part the Cherokee traders were quite shrewd, understanding well the value of various English goods and deerskins, and they frequently caught colonial agents in various underhanded schemes. They consistently complained, for example, of watered-down rum and inaccurate scales (deerskins were often purchased by weight rather than by piece).[38] In an effort to reduce this sort of corruption and ensure a regular flow of deerskins, in 1751 "South Carolina divided the Cherokee Nation into thirteen hunting districts, each of which 'should contain about two hundred gun men and three towns . . . in charge of one reputable trader.'" While this regulation more directly involved the colonial government in the fur trade, there is little indication that it eliminated or even reduced corrupt trading practices.[39]

Corruption was perhaps so prevalent because of the lucrative nature of the trade. There were great hazards in transporting various goods from coastal ports into the mountains, on poor roads and trails and through a landscape of shifting polities, followed by the equally uncertain return trip with a semi-perishable load of hides, but enough traders made significant profits to stimulate a great deal of envy and emulation (from both private and public interests). Prices paid for deerskins varied widely over time and locale, but relatively small quantities of European or colonial goods often bought traders substantial numbers of hides. Whereas a gallon of rum brought four high-quality skins, a musket could be exchanged for as many as thirty-five. In the early eighteenth century, South Carolina traders noted that just twelve flints or thirty bullets could purchase a good buckskin. A pistol would buy twenty, and a shirt would buy five.[40] In North Carolina in the early eighteenth century, a buckskin was valued in pounds sterling at two shillings and a doeskin at a shilling and sixpence, and in Charleston a raw buckskin was worth five shillings, with a well-dressed one fetching even more. Cherokee traders often saw better rates of return for their skins than Creek and other southeastern hunters. The larger deerskins from the mountains were particularly desirable, and the Cherokee people had special trade status with South Carolina as part of that colony's efforts to solidify friendly relations.[41] Although reliable records no longer exist to calculate rates of return on trading ventures, by all accounts the business was potentially quite lucrative.

A number of British colonists recognized the profitability of the hide trade and what those profits could mean for other aspects of the southern colonies' economies. The growing deerskin trade was not an activity separate from or lesser than commodity plantation agriculture. In reality the two were often intimately connected. Prominent planters tended to dominate the fur trade, even as smaller traders operated throughout the mountains, and commerce in hides was one of the potential avenues that a poor individual could follow to build wealth and enter the planter class. In Virginia the most prominent figure in the early fur trade was William Byrd I, one of the colony's most influential planters (see fig. 1.3).[42] In South Carolina the connections were even firmer. There a number of wealthy rice planters used their political standing to secure a share of the hide trade. Poorer men engaged in riskier trades and smaller deals in the hopes of elevating their social standing. Most failed, but a few succeeded and became important figures in their own right. Traders turned planters included a number of French Huguenot colonists, including John Guerard, Samuel Wragg, and Gabriel Manigault, who became political figures in part thanks to wealth built in the Indian trade. The capital flows were a two-way street, as rice and indigo plantations turned profits that enabled planters to finance trading expeditions, and the money made from deerskins was then funneled back into the agricultural economy, often used to purchase additional slaves. At a moment in time when South Carolina's fledgling plantation kingdom was putting down strong roots, the Indian trade furnished critical capital.[43]

FIGURE 1.3. Prominent planter and politician William Byrd I controlled a portion of the Virginia deerskin trade with the Cherokee towns during the early eighteenth century. Carol Highsmith photograph (2007) of Paul Cadmus painting (1939). Courtesy of the Carol M. Highsmith Archive, Library of Congress, Prints and Photographs Division.

Not only were the fur trading frontier and the plantation districts intimately connected, it can be argued that they were codependent. Historian Sven Beckert asserts that such was the case. Beckert argues that the use of commerce, violence, enslavement, and ultimately depopulation (what he terms "war capitalism") to secure land, labor, and natural resources in the colonial South was a crucial foundation for the plantation capitalism that was to spread over much of the Deep South in the nineteenth century. Along with trade in sassafras, ginseng, and Native American slaves, commerce in deerskins served as a leading wedge in more thorough appropriations of land for what would become the United States. Closing the loop, profits from these trades flowed into the pockets of large colonial planters and European investors, who would then invest them in African slaves, plantation lands, and ultimately cotton factories. In this recasting, a trade long portrayed as a rustic, adventurous, frontier endeavor becomes an account of sociopolitical domination, markets, and global capitalism. It connected the fur frontier to King Cotton.[44]

This expanding world of global commerce drew new participants into the deerskin trade and placed additional pressure on whitetails. By the mid-1730s, the newly founded Georgia colony had agents working the southern end of the mountain chain. During the 1740s this commerce expanded, and a deerskin-trading path leading down the Savannah River Valley through Augusta and to the port of Savannah joined the South Carolina and Virginia trade routes. Before Georgia abolished its ban on slavery in 1750, and as its colonists were still searching for a commodity crop, the fur trade helped drive its economic growth, much as had been the case in the early years of South Carolina.[45] This trade grew steadily through the following decades. In a period of eighteen years beginning in 1755, Savannah merchants would export approximately six hundred thousand deerskins to England.[46] Although Georgia traders would never handle as many hides as those from Charleston, they placed an even greater demand on deer populations at a crucial moment in the trade.

A confluence of factors—rising European demand for buckskin, new and more efficient firearms, expanding Indian demand for rum, Georgia's entrance into the fur trade, and sustained government efforts to make commerce in skins more efficient—made the 1740s and 1750s the peak years of the fur trade in the Southeast. By this time Cherokee society had largely reoriented itself and changed to suit the demands of commercial deer hunting. Firearms and solitary hunting had replaced fire, bows and arrows, and group hunts. European cookware and cloth had become important components of daily life in almost every town, and alcohol was an essential trade good, driving credit and debt relationships even as it often wrecked individual lives. Some towns had disappeared and others consolidated in the wake of the 1738 smallpox epidemic or because of the organization and location of colonial trading agents, who wielded inordinate economic power. Intermittent warfare with the Creek towns during this period also led to much movement and abandonment of historic sites of habitation.[47] What was unsettling for Cherokee people proved quite profitable for English traders. By the late 1750s Cherokee hunters alone were taking an estimated 150,000 whitetails each year, part of the much larger southern hide trade that was peaking at the same time. Even the colony of North Carolina, sandwiched between the more important exporters in South Carolina and Virginia and lacking proximity to the Creek and Chickasaw lands of the Deep South, exported thirty thousand deer hides in 1753.[48] Although rice had become Charleston's most important export, in the late 1740s the value of exported deerskins still "roughly equaled the combined total of indigo, beef and pork, lumber, and naval stores."[49]

From traders' perspectives, commerce in skins had always been about turning a profit, but the fur trade had long served political purposes as well. South Carolina in particular had attempted to expand or contract trade to reward or punish Native American groups, and at key moments like the Yamasee War it had leveraged its economic influence to create alliances. Trade had served as the primary tool in binding the Cherokee towns to South Carolina rather than to the Spanish or the French. This alliance threatened to dissolve in the late 1750s, however, in ways stemming from the fur trade and with grave repercussions for it. The event that would bring these tensions to a head was the French and Indian War.

From the moment fighting began in the North American theater of the Seven Years' War in 1754, the French encouraged and pressured the Cherokee governments to turn on the British. The encouragement included trading enticements, while the pressure entailed raids conducted deep into contested Kentucky deer hunting grounds by France's Shawnee allies. Although the English were unquestionably positioned to provide the Cherokee towns with more lucrative commercial opportunities, Cherokee leaders also realized that the colonial settlers trickling into the mountains by mid-century posed a grave threat, which surely must have made French entreaties more appealing. Part of the British effort to ensure a continued alliance with the Cherokee people was a pledge to build a series of forts in Indian territory to defend against Shawnee raids. In 1756 these plans were realized; South Carolina began construction of Fort Prince George on the upper Savannah River and Fort Loudon near Tellico, and Virginia erected a fort near Chota on the Little Tennessee River (though it was soon abandoned) as well as Fort Robinson on the Holston River.[50]

The purpose of these forts was twofold: they were intended to secure Cherokee loyalty to the British cause and to maintain the regular flow of deerskins to Williamsburg and Charleston. Some officials doubted their ability to ensure either goal, however. One of these was the chief engineer of Fort Loudon, John William Gerard de Brahm. A brilliant if eccentric man, de Brahm complained in writing to his South Carolina superiors that the fort would be ineffective since it was so distant from South Carolina's other garrisons and since supply routes were too tenuous for it to be able to withstand a siege for long. In other words, he warned that Appalachian geography would make British plans untenable. Disgusted when his warnings fell on deaf ears, de Brahm abandoned his work, and another military officer oversaw completion of the fort by the summer of 1757.[51]

De Brahm's fears proved prescient. In 1758 the Cherokee and English alliance fell apart, primarily because of a series of violent attacks along the Virginia frontier, due to a misidentification of Cherokee warriors by frontier settlers and subsequent retaliatory violence. Feeling betrayed and believing that the greatest threat to their sovereignty came from the inexorable westward expansion of the British colonies, Cherokee towns began open hostilities against the British in 1759 and turned on the colonial forts. As de Brahm had predicted, Fort Loudon proved to be impossible for South Carolina to resupply, and a Cherokee siege was successful. After eating their horses and dogs in an effort to hold out, the two-hundred-man garrison surrendered on August 9, 1760. The Cherokee leaders permitted the British troops to depart the fort, but, finding a cache of hidden arms and believing the surrender terms had been broken, they pursued the defeated soldiers and killed more than two dozen, taking others to ransom back to South Carolina.[52]

The violence culminating in the fall of Fort Loudon and attacks on the other frontier forts temporarily halted the regional fur trade and initiated the Anglo-Cherokee War, a regional conflict within the larger French and Indian War. The colonial response to the fall of Fort Loudon was rapid and harsh. Britain sent regular soldiers as well as South Carolina militia forces into the mountains on campaigns to destroy Cherokee towns and crush Indian military opposition in late 1760 and again in 1761. By all accounts the campaign was a hard one, with civilians as well as warriors killed. Militiamen targeted the Cherokee agricultural infrastructure in particular, making a special effort to burn cornfields wherever they found them, and wrought widespread havoc in the Lower and Middle Towns.[53] Facing these British tactics the Cherokee towns sued for peace, and with many of their town sites in shambles and grain supplies burned, hunting whitetails was of more economic importance than ever before. Famed botanist William Bartram, traveling through Cherokee territory before the peace had been finalized, found frequent evidence of hide hunters and encountered fur traders in almost every town he visited despite the hazards.[54]

Although the deerskin trade, supplemented by commerce in other fur-bearing mammals, had driven Appalachian economies for several decades, and the end of the Anglo-Cherokee War seemed to restore the trading status quo, it would not last indefinitely. The trade in hides on the scale Bartram witnessed would soon be a thing of the past. The 1760s marked a serious decline in the deerskin trade for Cherokee people, and the commerce would be all but finished by the time of the American Revolution. This decreasing trade came from a confluence of related forces. By some accounts the number of white-tailed deer in the southern mountains plummeted after the war, Cherokee political and military power declined (and with it their control over hunting territory), and the encroachment of colonial British and then American settlers further disrupted the old patterns of hunting and trade.

Southern whitetail numbers had continued their natural fluctuation after the initiation of transatlantic commerce in hides in the late seventeenth century, but the overall trend was toward declining herds as the number of deer killed skyrocketed in the early 1700s. One place to see concern about dwindling deer populations is in early colonial hunting regulations. These statutes restricted deer hunting in some manner, often by limiting the times of year when deer could be taken, in an effort to ensure a continued supply of whitetails. These laws were not driven by a belief in the importance of conservation or equity for Native peoples who depended on deer for sustenance, but rather were intend to maintain an important economic commodity. Virginia would regulate deer hunting as early as 1699, followed by North Carolina in 1738, South Carolina in 1755, and Georgia in 1790. These laws reflected real or perceived regional declines.[55]

Although the overall trend was downward, evidence suggests that the collapse of whitetail populations was uneven in ways that connected to both ongoing environmental fluctuation and the dynamics of human politics and trade. Years with heavy mast crops and particularly good weather could produce many fawns. Local political conditions could also create pockets of dense deer populations even as stocks of the animals declined in other areas. The Yamasee War, for example, led to a period of slow deerskin trade, with a likely rebound of deer populations as a consequence. Likewise, border regions where territories of two warring polities met could create a sort of no-man's-land where hunting was dangerous and deer thrived. This occurred in pockets of the Deep South from the 1730s to the 1750s as the Choctaw and Chickasaw peoples engaged in a series of conflicts with one another, and there is evidence of a similar phenomenon in Appalachia.[56] The Anglo-Cherokee War seems to have led to a brief rebound of deer, for example. Henry Timberlake, writing near the end of the conflict, noted "the amazing quantity" of game he witnessed during and shortly after his campaigning.[57] Likewise, Christopher French, who participated in South Carolina's campaign against the Cherokee towns in 1760, saw numerous deer in the foothills of the mountains, which served as the middle ground between the colonists and Indians in the struggle.[58]

The French and Indian War, in which the Cherokee towns first sided with and then fought against the English, put a temporary halt to the Appalachian deerskin trade, but its graver lasting consequence was the British victory in the larger conflict. During the 1750s colonists had begun to push into the eastern edges of the mountains, especially in Virginia and North Carolina, and although the crown imposed prohibitions on settlement west of the crest of the Blue Ridge Mountains in 1763, these were largely ignored and eventually modified and then repealed. The 1760s saw an influx of white settlers moving into Cherokee hunting territory. These new arrivals cut timber and cleared small patches of farmland, hunted whitetails for food and for skins, and clamored for the colonial government's protection from Native Americans. The "long hunters," as some woodsmen among them came to be known, ventured even farther into Cherokee territory and the hunting grounds of Kentucky that lay between the Cherokee towns and Shawnee land. The Shawnee people added to this encroachment pressure as well, venturing farther south more frequently, as they too felt the pressures of the fur trade and settlers from the north and east. These combined forces, coupled with the recent war's devastation, seem to have caused Cherokee hunters to contract their hunting territories, pursuing deer closer to the relative safety of their towns and rendering some of the more distant yet productive hunting grounds no longer accessible.[59]

As early as the 1740s, Cherokee leaders formally complained to Georgia's governor about white hunters taking deer on their lands, and complaints to representatives of the other southeastern colonies came more frequently as the century progressed. For their part, white settlers believed taking deer to be their right, part of the process of taming a wilderness and making a living. Many seemed to believe, as Timberlake wrote in 1765, that "the sole occupations of an Indian life, are hunting, and warring abroad, and lazying at home."[60] According to this line of thinking, it was white settlers who would improve the land and make it a productive part of the colonial empire. Although the majority of frontier settlers engaged in solitary, local hunting for subsistence purposes, an increasing number of organized hunting groups engaged in more systematic participation in the deerskin trade. These parties undertook extended journeys into traditional Cherokee hunting grounds like southern Kentucky and saw these endeavors as their right, as did an organization of approximately fifty Virginians who threatened violence to any Cherokee hunters who attempted to interfere with their large-scale participation in the hide trade in the 1760s. After the Anglo-Cherokee War, deerskins sold by colonial settlers in the mountains to South Carolina traders exceeded those sold by the Cherokee for the first time.[61]

These settlers also brought livestock that added to the challenges for mountain deer populations, as each year increasing herds of free-ranging cattle and hogs ate up mountain mast and browse. By the 1760s the Cherokees themselves had fully embraced raising hogs and were beginning to adopt cattle as well. This turn to livestock may have reduced hunting pressure on deer for sustenance, but it likely more than made up for it by altering the habitats on which deer relied. Livestock's competition with deer for grazing resources would only increase in the years to follow.[62]

The downward spiral of the deerskin trade was largely completed by the time of the American Revolution. Although Cherokee hunters would continue to take deer and sell hides in much smaller numbers into the early nineteenth century, the colonists' victory removed the last prohibitions on frontier settlement and flooded the region with farmers intent on changing Appalachian landscapes and converting the mountains to private property. Exactly how many Appalachian whitetails became a part of the transatlantic trade is all but impossible to determine, but it was certainly a significant number. In any given year between 1700 and the Revolution, Cherokee hunters took tens of thousands or even hundreds of thousands of deer, for a total of several million above and beyond the needs of subsistence and local economies, and Virginia and Charleston traders exported approximately six and a half million deer hides to Europe. By the 1780s, there were already pockets of the southern mountains in which whitetails had been all but extirpated. American victory and subsequent new waves of settlers streaming into the mountains merely capped two decades of fur trading decline. Furs would continue to be bought and sold in East Tennessee well into the nineteenth century, but the deerskin trade as a driving economic engine was no more. As historian W. Neil Franklin wrote in 1932, the war virtually "brought an end to a trade, the collapse of which could not have been long delayed."[63]

శ్రీ

During the nineteenth century, memory of the Appalachian fur trade and its nodes like Fort Loudon faded to the realm of myth and set pieces in romantic stories of frontier settlement. Representative was their incorporation into a Victorian novel—*The Story of Old Fort Loudon*—by one of the most

"The men had been hastily formed into a square."

FIGURE 1.4. This sketch of combat outside Fort Loudon during the Anglo-Cherokee War illustrated Mary Noailles Murfree's romantic interpretation of the Appalachian fur trade's dying days. Illustration by Ernest C. Peixotto, in Charles Egbert Craddock (Mary Noailles Murfree), *The Story of Old Fort Loudon* (New York: Macmillan, 1899).

influential literary "discoverers" of Appalachia, Mary Noailles Murfree (who wrote under the pen name Charles Egbert Craddock). Murfree's protagonists venture into the mountains of East Tennessee during the Anglo-Cherokee War and at various points feign being fur traders and French to trick Cherokee warriors hostile to English colonists, only to witness some of the brutality of the conflict that resulted in the fort's downfall.[64]

Today a reconstructed Fort Loudon once again overlooks the Little Tennessee River Valley, although the vista is quite different than it was in the late 1750s. A large reservoir, one of many such impoundments created across the region in the twentieth century by the Tennessee Valley Authority (TVA), stretches across the valley, and tourists intent on outdoor recreation or historical tours stream by the fort. Roughly a quarter million visitors walk through the gates of the fort each year to learn about colonial American expansion, the Cherokee people, and the French and Indian War through programs managed by the Tennessee Department of Education.[65]

This modern Fort Loudon hardly represents a contemporary appreciation for the power and influence of the eighteenth-century deerskin trade, however. It is much more a product of late-twentieth-century politics. Although the fort site was named a National Historic Landmark in 1965, it was the development of the TVA's Tellico Dam and the opposition it drew in the 1970s that brought attention to Fort Loudon's historical importance. The site was slated to be flooded by the new dam and made for a handy tool in the legal struggle over the construction. In particular, the Eastern Band of the Cherokee Indians—remnant unmoved survivors of the struggles over land and resources that had once played out in places like the Little Tennessee Valley—challenged the erasure of historic sites such as Fort Loudon for the sake of questionable economic and recreational benefits. They especially challenged the flooding and virtual erasure of the site at Chota, historically a center of tribal government and religion, which one Cherokee representative called "the most important place in all of Cherokee life."[66]

The controversy that surrounded the TVA project initiated a great deal of research. Archaeologists, with the support of the agency, excavated the fort site and learned a great deal about its construction and material culture, and they conducted similar studies at a number of known Cherokee settlement sites within the proposed reservoir's footprint. This scientific work, although it unearthed many artifacts affirming the long history and importance of several Cherokee sites, also laid the groundwork for the reservoir. Along with a court ruling that denied that the drowning of the Little Tennessee Valley would deprive Cherokee people of their guarantee of free expression of religion through loss of access to sites like Chota, it spelled the end of the legal battle against the dam, as the government argued that it had made good faith efforts to document the historical and cultural significance of the region prior to inundation. As a concession, the TVA also pledged that the Loudon site, for a few brief years so central to politics and trade in southern Appalachia, would remain accessible even after the dam closed and the reservoir filled. Construction workers used 250,000 cubic yards of fill dirt to raise the entire site approximately twenty-five feet above its old elevation and safely beyond the reach of the reservoir. On this new landscape archaeologists reconstructed Fort Loudon with materials that "were sympathetic to the originals," and the TVA paid for a modern visitors' center, complete with museum exhibits. This re-created complex now commands a view of Tellico Lake and its sunbathers, fishers, and jet-skiers.[67]

The TVA's new Fort Loudon proved much more effective than the original one constructed by the colony of South Carolina. It successfully quelled opposition to the Tellico Dam, as the archaeological work conducted throughout the valley and the TVA's quick responses to Cherokee demands quieted many tribal opponents, while the new Fort Loudon Historic Area centered on the rebuilt fort took some of the punch from historic preservation groups intent on blocking the dam. As two historians of the struggle over the dam's construction observed, agency officials sagely believed that "cooperation with the Eastern Cherokees would reap tremendous public relations benefits for TVA."[68] They were right. The TVA seemed to agree with John H. DeWitt, who gave an address upon the dedication of a monument at the site of Fort Loudon in 1917 when he was president of the Tennessee Historical Society. "The story of old Fort Loudon has naturally been invested with romantic and melancholy interest," DeWitt had said, but "the enmities and rivalries which caused the erection and then the destruction of Fort Loudon have long since disappeared."[69] In their place, the TVA argued, was a new spirit of cooperation and progress.

Neither DeWitt nor the TVA acknowledged that Fort Loudon's significance resided less in its brief and inglorious military life than in its symbolism as a physical manifestation of the trade in deerskins that drove commerce and politics in the southern Appalachian Mountains (as well as substantial portions of the lowland South) for nearly a century. Deer as much as anything else had lured colonial merchants into the mountains, shaped diplomatic relations between the seaboard colonies and the Cherokee towns, and altered patterns of hunting and subsistence over thousands of square miles. The trade had helped spread contagious diseases, diffused firearms, changed forest composition, and contributed to tensions among European powers and Native American people. It had also taxed some whitetail populations to the breaking point and transformed parts of the anatomy of a wild species into a commodity.

What if anything survived the demise of the regional fur trade? Perhaps the most important outcome was a pattern of viewing the southern highlands as a region that produced valuable commodities (indeed, the mountains were akin to the entire American West in this regard). Deerskins would prove an early example of a repeated phenomenon in the Appalachian Mountains. The environment, rather than isolating mountain residents as soon-to-emerge stereotypes in popular culture would suggest, provided a range of natural resources that fostered a variety of connections to other regions. Appalachian nature would prove to be more a unifying force than an isolating one, and the story of deerskins would be repeated, on varying scales and with differing details, time and again over the two centuries to follow.

CHAPTER 2

Plants

Botanical Collectors and the Roots of Appalachian Identity

RETURNING TO HIS COLLECTING BASE near the town of Asheville in the mountains of western North Carolina in the summer of 1804, botanist and plant collector John Lyon must have presented quite a sight. His horse would have carried bundles of rootstock, sacks of seeds, wooden presses for drying plant specimens, sheaves of field notes, and the various provisions necessary for roaming the mountains for weeks on end. Lyon himself certainly would have been dirty and worn from his journeys, having endured months of long rides on rough mountain trails, camping out in booming thunderstorms, and bushwhacking up narrow coves and rocky ridges in pursuit of rare plant specimens. To complete the scene, picture the collector suffering grievously from a spectacular skin rash. It was an affliction he believed came from collecting the seeds of Michaux's sumac (named for famed botanizer André Michaux) but likely was instead the result of a misidentification of poison sumac.[1]

For all the interest of this scene, Lyon was hardly a unique figure in early-Republic Appalachia, especially in the mountains of western North Carolina, where certain spots, such as Roan Mountain and Grandfather Mountain, became meccas for traveling scientists. The southern mountains were important collecting grounds for American and foreign botanists in the late eighteenth and early nineteenth centuries, and the same botanists who collected species new to Western science also frequently recorded their thoughts on Appalachian environments and people. These accounts were not solely for private amusement; they reached and influenced a larger audience. Collectors' writings came not just out of dispassionate observation and rumination, since they stemmed, at least in part, from a direct economic interest. Botanizers (to use a term then current for professional botanical collectors) made a living from selling plants in multiple forms. They dealt in dried and live specimens and seeds, they won government and private positions based on their botanical expertise, and some sold their travel narratives as books. They traded in a commodity made valuable by its exclusivity, and thus they had powerful incentives to represent

Appalachia as exotic and isolated. Emphasizing the region's wildness added value to their specimens and knowledge and made for juicier travel stories. Portraying Appalachian people as backward and rustic only added to the idea that the mountains were isolated by nature.

Appalachian exoticism, the idea that the southern mountains are particularly remote, backward, and insular, is often associated with the cultural "discovery" of Appalachia in the late nineteenth and early twentieth centuries. Some of the most prominent "discoverers" included Berea College president William Goodell Frost (who declared in 1899 that eastern Kentuckians had been so isolated that they constituted Anglo-Saxon Americans' "contemporary ancestors"), geographer Ellen Churchill Semple, and regional novelists such as Mary Noailles Murfree and John Fox Jr.[2] While cultural critics did much to popularize the idea of Appalachian exoticism, they drew on a long tradition of such portrayals. Largely forgotten, but of great importance in the creation of this image, were the characterizations of botanists of a century prior. Botanizers' writings lived a second life, as their work was read a century later by a range of Appalachian identity makers, including historians, travel writers, and novelists. There is solid evidence that these "discoverers" of Appalachia were influenced by the work of botanical collectors. Reading these botanists as the first wave of Appalachian identity makers provides a fascinating prologue to the classic late-nineteenth-century and early-twentieth-century discovery of Appalachia and roots it in the pursuit of profit from an early regional commodity: plant specimens.

Historians of the region have acknowledged the presence of these collectors in the southern mountains, but they have rarely presented an overarching analysis of collectors' actions, voices, or importance. Botanizers most frequently appear in these studies as tourists with singular voices, interesting if anomalous individuals, discoverers of the new and the rare. Even more frequently they are portrayed as witnesses of Appalachian nature and culture at the dawn of regional Euro-American settlement or shortly thereafter. They describe nature or Indian life (or both) during the time "before," soon to be transformed into a degraded landscape of "after." Put another way, they write of walking through Eden just before the fall. Botanizers were also, however, members of a burgeoning profession—scientist—in which individual actions and observations were tied to larger economies and worldviews. This professional gaze would shape Appalachia greatly at the end of the nineteenth century, but it was not then an entirely new interpretive force. Another untold story is the way in which later professionals—scientists, historians, and novelists among them— heavily drew on the interpretations created by these early botanists. A closer look at these men and their work reminds us that as a class of goods the rare plants themselves, as well as knowledge about their ranges and life cycles, functioned as a sort of commodity.[3]

The work of botanical collectors accomplished more than moving plant materials around the globe; it also transferred ideas. Botanists brought outside eyes to the southern mountains, and collecting expeditions often resulted in descriptive accounts of Appalachian residents and places in addition to new entries in taxonomic texts. More than just elitist observations of local people, these accounts were some of the earliest and most repeated characterizations of life and culture in the southern mountains. They helped build a picture of regional isolation and disconnection rooted, ironically, in an endeavor dependent on burgeoning global interconnection: the international plant trade. Even as they sold little bits of Appalachian nature, botanizers used that same nature to sell particular stories about the southern mountains. These plant collectors would begin to construct an image of Appalachian otherness, a portrayal of people and place that would dominate American ideas about the region by the end of the nineteenth century.

ᥱᴖ

Appalachia's apparent wildness did much to lure its first botanizers. Much of the southern mountain region had been settled by Euro-Americans, at least sparsely, by the late-eighteenth century, but pockets of the mountains remained somewhat remote, with more relatively undisturbed native vegetation (much of it awaiting inclusion in scientific taxonomy) than most landscapes in the eastern lowlands. In addition, the mountain topography aided plant diversity. Thanks to varied elevations, microclimates (created by sheltered valleys, rain shadows, and topographic uplift), and a wide range of soil types, the southern mountains supported an astonishing number of ecosystems, often in close proximity. Each ecosystem held distinctive plant and animal life, furnishing collectors with additional opportunities to harvest desirable specimens or even describe species new to Western science.

A roster of prominent botanists who roamed southern Appalachia reads like a Who's Who of late-eighteenth- and early-nineteenth-century science. Such luminaries as John and William Bartram, André and François André Michaux, John Fraser, John Lyon, Frederick Pursh, Matthias Kin, Thomas Nuttall, Constantine Rafinesque, Moses Curtis, Samuel B. Buckley, Asa Gray, and Leo Lesquereaux frequented the region.[4] These botanizers were more than just curious men of science (universally male); they were also practitioners of another early Appalachian extractive industry. Traffic in plant specimens mimicked the trade in deerskins, if on a smaller scale, and connected wild nature, isolated people, and economic value in ways that would be echoed in later regional descriptions.

ಌ

Although botanists sought rare plants, their travels often had much in common. First, collectors followed regular routes of sorts—the "trade paths" of botanizing—although they were always willing to take a side trail in search of undiscovered specimens. One prominent path through the mountains followed the wagon ruts of early mountain settlers, traversing the Great Valley from Pennsylvania southwest through Virginia and into Tennessee and accessing the high peaks of the Blue Ridge and the Smokies from the west. Another equally popular route began in either Charleston or Savannah and ascended the rivers through the Piedmont country into the mountains, often with either Asheville or Morganton, North Carolina, serving as a base for collecting forays to high peaks like the Black Mountains or the Roan highlands. A third trail crossed the Alleghenies in southwestern Pennsylvania with a stop at Pittsburgh, then descended the Ohio River, offering opportunities for collecting excursions along either bank and access to the dissected plateau of the Cumberlands in eastern Kentucky and east-central Tennessee. From there botanists might travel on to the Gulf South or the trans-Mississippi frontier.

These paths had regular stops, as certain households became way stations on botanizing routes. For example, Waightstill Avery, living near Morganton, provided board to André Michaux, John Fraser, Moses Curtis, and other collectors.[5] John Lyon liked to stay with the Erwin family when in the vicinity of Asheville.[6] And almost every collector of note who passed through northwestern North Carolina on either side of the turn of the century stopped at the home of Martin Davenport, situated on the bank of the Doe River along Bright's Trace, the main road between western North Carolina and the settlements of northeastern Tennessee.[7] These homesteads offered more than a bed, food, and advice. Botanizers heavily relied on local guides to assist them in exploring more inaccessible reaches of the mountains. Examples are numerous: Davenport himself guided André Michaux up Grandfather Mountain (Michaux had earlier employed an Indian guide to explore the Black Mountains to the south), a guide helped Lyon descend Roan Mountain when rough weather set in, Constantine Rafinesque relied on a professional hunter to lead him to the falls of the Cumberland River, Silas McDowell recalled regularly assisting Lyon on his visits to western North Carolina, and Billy White led Curtis in multiple ascents of Grandfather.[8]

These frequent visits, the formation of lodging and guide networks, and established collecting circuits came from the value rare plants held in transatlantic commerce of the day. Plant collecting was serious business during the early Republic. Wealthy European enthusiasts had been engaged in collecting natural specimens of all types for more than two centuries by that time, first for private collections, often styled "cabinets of curiosity," and then in more formal private and public museums. This collecting impulse accelerated with the height of European colonial empires and global voyages of discovery, as rare plants and animals flowed almost as freely as people, commodity crops, and diseases across the world's oceans.[9] The peak of colonial natural history collecting may have been the late eighteenth and early nineteenth centuries, as collectors and scientists attempted to inventory the world's biological diversity and come to grips with the relationships between species and their global distributions.[10] Centers of botanical knowledge and economic activity, such as Paris's Jardin des Plantes and London's Royal Botanical Gardens at Kew, amassed enormous collections of living flora, dried herbarium specimens, and seeds, using these stores for research and aesthetic purposes and to support intra-empire plant transfers with economic potential.[11] Botanizers roamed the four corners of the earth throughout the late eighteenth and early nineteenth centuries, searching for plants that were new to science as well as species with ready markets in the metropoles. Within these colonial contexts plants were power, and this provided great motivation for botanists to stress the hazards of their travels and the remoteness of collecting sites.

Some collectors worked on commission from wealthy patrons. William Bartram was a prominent example. From 1772 to 1776, English collector John Fothergill provided him an annual stipend and paid for botanical illustrations made from Bartram's tour of the southern colonies.[12] Others, such as Michaux and Fraser, served as official scientific agents of empire, working for various European governments, while Lyon and some fellow botanizers collected mainly for personal aggrandizement. For example, Lyon's collecting endeavors were twofold. First, he managed Woodlands, a botanical garden in Philadelphia, around the turn of the century, and he was responsible for adding to its collection. (Management of Woodlands passed to another well-known collector, Polish-born Frederick Pursh, after Lyon left the garden in 1802.) Second, he sold specimens on his own account to wealthy collectors.[13] A journal Lyon kept illustrates the careful—but typical—calculations that were part of botanizing. For each trip into the mountains he duly noted mileage, the expense of various services, and the costs of horse husbandry. Lyon also noted sales of resulting specimens in places like Philadelphia and all of the associated costs involved in preserving seed and rootstock, cultivating seedlings, shipping plants for sale in England, and he kept track at length of the various expenses involved in arranging overseas auctions.[14]

It may be true that thanks to Lyon, as a botanist of his day argued, "more rare and novel plants have been introduced from thence (North America) to Europe than through any other channel whatever," but his work also gained monetary rewards.[15] In one auction sale of Appalachian specimens on the British market, Lyon determined his total expenses to have been a bit over 668 pounds sterling, but the transactions netted him a profit of more than 613 pounds.[16] A popular twenty-first-century science writer estimates that a two-year span of collecting garnered Lyon "almost $200,000 in today's money for his efforts."[17]

For a time, André Michaux's business model was quite similar. He founded a botanical nursery in northern New Jersey and shipped an estimated sixty thousand plant samples to France between 1785 and 1793. He was particularly interested in the commercial uses of North American hardwood species, which the French navy considered a potential shipbuilding resource.[18] John Fraser also sold plants on his own account, in addition to providing rarities to the Russian royal court. Fraser published lists advertising selected specimens to English collectors who might visit his Chelsea home. One such catalog invited "Ladies and Gentlemen who will do him the Honour to call and see them" and included a long list of rare or showy plants, many of which—like the "most beautiful plant," *Rhododendron punctatum*—came from southern Appalachia.[19] Another of Fraser's moneymaking schemes involved collaboration with South Carolina planter Thomas Walter to collect and sell upland bentgrass seed in Europe for "two guineas a quart," although this endeavor never proved profitable.[20]

Later botanizers like Asa Gray rarely sold specimens, but discovering and collecting plants new to science still carried financial rewards in the form of career advancement. In Gray's case his travels aided his ascendance from the faculty of the University of Michigan to the Fisher Professorship of Natural History at Harvard University.[21] Constantine Rafinesque sold knowledge too, drawing on his expertise in plants (as well as ichthyology and geology) to secure a post at Kentucky's Transylvania College, and he used descriptions of new biological discoveries in the cornerstone articles in his own periodical, the *Atlantic Journal, and Friend of Knowledge*, launched in the early 1830s.[22] The field also contained more than a few bi-vocational amateurs who derived pleasure from mastering the flora of their vicinity and exchanging plants and correspondence with botanists from other places. A classic example is George Engelmann, a St. Louis doctor who devoted his non-work hours to botanical studies and who built and maintained connections with such scientific lumi-

FIGURE 2.1. Asa Gray, between 1855 and 1865. Courtesy of the
Library of Congress.

naries as Curtis, Gray, Nuttall, and British authority William Hooker. Engelmann sought intellectual company but also mountain specimens, such as Fraser's sedge ("a very interesting rarity"), from his Appalachian correspondents.[23]

Whatever form the benefits of botanizing took, collecting was often a strenuous endeavor. Lyon, for example, periodically noted the challenges of botanizing in the mountains, constantly portraying Appalachian environments
as more rugged and taxing than other southern landscapes. Occasionally he

elaborated on his frequent notations of "roads rough and bad" and "Accommadations bad."[24] Caught in a mountain thunderstorm near Asheville in 1804, Lyon sought shelter for the night. The next morning he marveled at nature's power, noting, "A shocking scene of havock presents to the eye this morning from the effects of the late storm, roads in many places impassable."[25] Four years later on Roan Mountain, along the North Carolina and Tennessee line, he was again caught in a storm when heavy fog rolled in. Lyon's party found their way down the mountain, and he wrote that they were fortunate to have not "perished before the storm was over with cold and hunger."[26]

Potentially deadly landscapes often contained divine elements too. When Lyon climbed the divide between the Keowee and French Broad Rivers in southwestern North Carolina during an 1807 collecting trip, he described the landscape as "in many places so steep and rugged that a Horse can hardly clamber up indeed it is in some places no more then 30 or 40 yards wide, while the view on each side to look down is frightfull, but the distant view is sublime."[27] William Bartram was perhaps the most effusive in his praise of mountain scenery, including passage after florid passage describing the scenic qualities of southern Appalachia, especially in sections where travel proved "very rough and troublesome."[28] But Bartram understood the dangers of the mountains, confessing in 1775 that while the Appalachians offered the promise of "some new things," "it's look'd upon as hazardous."[29] And Sicilian-born Rafinesque found the Alleghenies of southwestern Pennsylvania rough going, yet of an 1819 trip he wrote that he traveled the mountains "on foot, as I never would cross these beautiful mountains in any other way, in order to botanize all the while, and I was rewarded by many new plants."[30]

In account after account, the ecstasies and challenges of remote mountain excursions were intimately intertwined. André Michaux's first ascent of Grandfather Mountain, in present-day Avery County, North Carolina, highlighted the elation collectors sometimes felt when conquering the challenges of an expedition. After a strenuous climb of the rocky peak, accompanied by guide Martin Davenport, bushwhacking through the dense undergrowth only to emerge into the clear, clean air of the summit, Michaux must have been overcome by emotion. He wrote in his diary of singing "La Marseillaise" and chanting "Vivre l'Amérique et la Républiq. Francaise, Vive la Liberté &c &c."[31] It was this sort of transcendental experience that Scot Alexander Garden (namesake of the gardenia) so regretted missing when his North American travels were cut short, precluding a journey through the southern mountains. Garden bitterly lamented being denied the "prospect of glutting my very soul with the view of the southern parts of the Great Apalachees."[32]

Other botanizers too described Appalachia as a beautiful but hazardous place. François André Michaux, traveling in southwestern Pennsylvania, wrote of an endless set of ridges rising up before him, "present[ing] a picture resembling the sea after a storm," and he noted with morbid curiosity tales of rattlesnake attacks.[33] His father, André, had been even more explicit concerning the difficulties of southern mountain travel; his journal was full of accounts of trips that wore down men and horses.[34] Rafinesque also advised aspiring collectors of the hazards of working in the wilds of eastern America, warning of pesky insects, challenging weather, long periods of solitude, and "rough or muddy roads to vex you, and blind paths to perplex you, rocks, mountains and steep ascents. . . . You may be lamed in climbing rocks for plants, or break your limbs by a fall."[35] Be that as it may, "The pleasures of a botanical exploration fully compensate for the miseries and dangers; else no one would be a travelling botanist."[36]

The elder Michaux's accounts of the strenuousness of Appalachian collecting so impressed one contemporary Irish translator that his summary of the botanist's travels dwelled repeatedly on the hazards of "these uninhabited countries [where] the forests are almost impenetrable, there being no other tracks than those formed by bears." According to this admirer, by force of necessity Michaux "was incessantly engaged in climbing rocks, or passing torrents; often upon the rotten trunks of trees, which crumbled beneath his feet; where a frightful darkness rests over the wilds, produced by the thickness of the branches interwoven with climbing plants, and still more by the almost continual fogs, which cover these rugged mountains."[37] William Hooker also noted that the Frenchman's work in the southern mountains was replete "with great difficulty and danger," slogging through beautiful but imposing "thickets of *Rhododendron*, *Kalmia*, and *Azalea*."[38]

Writers often used comparison with other, distant sites of botanizing to highlight the challenges of Appalachian collection. For example, Michaux traveled the world collecting for the French government, including work in the mountains of Persia, the deserts of Syria and Iraq, Spanish Florida, the windswept shores of Hudson Bay, Tenerife in the Canary Islands, and the tropical forests of Madagascar, where he finally lost his life to "the fever of the country."[39] With conceptions of disease at the time, when people were quick to attribute maladies to humoral imbalances or miasmas produced by nature, the Madagascar environment may have killed Michaux. But only in southern Appalachia had he found it imprudent to travel with only a single guide, "there being various accidents by which one may perish," leaving the collector alone in "woods impenetrable to an European."[40]

These seemingly exaggerated descriptions of the southern mountains served a practical purpose. The conclusion implicit in all these accounts was that the most challenging environments yielded the rarest and hence most valuable plants. The same Keowee climb that Lyon found so hazardous also resulted in a satisfied journal entry noting many "rare mountain plants."[41] John Fraser, describing the beauty and challenge of ascending Roan Mountain, emphasized that his prized specimen of showy *Rhododendron catawbiense* came from the peak.[42] Moses Curtis likewise made explicit the connection between scenic beauty—whether threatening or comforting—and rare species, as when he speculated upon approaching the Pigeon River gorge that, thanks to its hazards, it "must afford fine scenery and doubtless furnish some valuable plants."[43] André Michaux described the climb though the Cumberland Gap as quite strenuous, but at the divide he stopped to marvel at, and collect, "a climbing fern covering an area of over six acres" that was uninjured by an exceptionally severe frost.[44] And Rafinesque struggled to reach the upper stretches of the Cumberland River with the help of a guide but for his pains "was rewarded by many new and rare plants."[45]

Exceptionally rare plants, and the nature of their limited habitats, could stimulate imaginations and build careers, binding Appalachian climes to intellectual niches. The most famous case is that of *Shortia galacifolia* (now commonly known as Oconee bells), which linked the collecting of North America's two most famous botanizers: the elder Michaux and Gray. On a 1787 trip into the Appalachian range from the southeast to collect shrub and tree species that might sell well as ornaments to Europeans gardens, Michaux dug and preserved a specimen of a small herbaceous plant. More than a half century later, in 1839, Asa Gray was touring the collections of Paris's Jardin des Plantes and came across the dried sample. He instantly recognized that not only was the plant an un-described species, it seemed to belong to an entire genus hitherto unknown to science. Gray named the plant *Shortia galacifolia* and became obsessed with rediscovering it in the wild, completing one part of the famed Michaux's work in the southern mountains.[46]

The trouble was that Michaux left few clues about where he had found the plant. His collecting took him to almost every valley and peak of Appalachia, and the brief note accompanying the Jardin des Plantes' specimen said only that he had found it in the "high mountains of Carolina."[47] Because of its rarity, Gray became convinced that it must be from the most rugged and inaccessible reaches of the mountains, and he eventually fixated on the peaks of northwestern North Carolina in the vicinity of Roan and Yellow Mountains. Gray participated in several search expeditions to the region, sought *Shortia's* potential relatives in places as far afield as Japan, and regularly communicated with fellow collectors about the mysterious plant. Eventually his quixotic *Shortia* obsession became something of an amusing professional anecdote.[48]

Nearly forty years after the search began, in 1877, the thirteen-year-old son of another botanizer would rediscover a patch of *Shortia* in McDowell County, North Carolina, to the south of Gray's search zone. Gray would bring the collection to international attention in botanical publications, crowing (somewhat speciously) that it had been found "in the district I had indicated as the most probable locality."[49] Gray had been mistaken about *Shortia*'s preferred environmental niche: later botanists would discover that it grew best on wet stream-side slopes in the lower mountains, concentrated most thickly in the district where South Carolina, North Carolina, and Georgia come together. Michaux likely collected his sample along the Keowee River or where the Horsepasture and Toxaway Rivers join.[50] Gray may have been mistaken about *Shortia*'s habitat, but his quest epitomized the linkage of rugged mountain environments and valuable plants: it made perfect sense to the eminent scientist that Appalachia's most elusive plant must be found in some of its most rugged and remote terrain. Rarity and isolation were linked in his mind and shaped his career path. When an elderly Gray reached his seventy-fifth birthday, the editors of the *Botanical Gazette* sought a gift that would convey the scientific community's esteem for him and summarize his life's work. They settled on a commemorative trophy cup, paid for in part by Gray's fellow botanists, engraved with images of two of the many plant species that Gray had introduced to science. One of the species depicted was *Shortia galacifolia*.[51]

Collectively, Appalachian plants might be defined as a commodity, but as individual specimens—as the long hunt for a single example of *Shortia* demonstrates—they were most certainly luxury goods, with their value derived from their rarity and novelty. Wealthy collectors and botanical gardens sought showy, unique plants, and botanizers wanted to build their reputations and fortunes by discovering and describing new species or selling valuable ones. To be clear, this created a financial incentive to describe the southern mountains as a rugged if beautiful wilderness. Asserting the difficulty of collection could encourage higher prices from collectors who craved exclusive plants. Finding a specimen in a quotidian landscape, say a hedgerow, hardly made for good advertising. In the case of botanists like Michaux or Gray who rarely sold plants, stressing the challenging nature of their work could still pay off, as it emphasized the need for continued funding from the French government or an American university in their respective cases. Rafinesque summed up this connection between exoticism and perceived value succinctly in 1833. Even at this late date, Rafinesque described southern Appalachia as "the least known of all our mountains" and noted that this perceived obscurity offered great promise. He argued that the mountains undoubtedly held mineral wealth and observed that they already attracted speculators with "valuable mines of coal, iron, gold," but he thought their greatest riches were "the unexplored fossils, flowers, animals and precious stones which I know they contain."[52] As an added incentive to emphasize the difficulties of their work environments, unforgiving yet beautiful nature made for good narrative for botanists planning to publish their travelogues.

Discussion of the natural exoticism of Appalachia often extended to unflattering portrayals of its people. To some extent, elite and well-traveled botanists describing local farmers as backward and primitive is unsurprising. For as long as travelers the world over have recorded their journeys they have made such comments. But to dismiss these critiques as banal is to ignore the ties between these characterizations and the economic incentives of collecting. Portraying mountain people as nearly as wild as mountain environments made a certain financial sense. Describing mountaineers as rustic and insular—isolated—reemphasized that botanists collected from wild and hard-to-reach places, adding further value to their specimens and expertise.

Lyon, for example, periodically commented on the backwardness of people in the region, where it "be corn bread for man, and corn feed for Horses."[53] Visiting the famous Scots-Cherokee planter James Vann, he criticized one of North Georgia's wealthiest men as "a truly mad and dangerous animal," especially "when intoxicated," and praised the work of the local Moravian mission in reforming Cherokee and Euro-American locals.[54] Although this portrayal might have drawn on Lyon's prejudices concerning Native Americans, it was also consistent with his other comments on regional residents, as was the case at the Estatoe settlement (where the Carolinas and Georgia met) when he disparaged the people as "rather a rude set."[55] Occasionally Lyon expressed surprise at less "rude" residents, as upon encountering a "sensible intelligent man considerably skilled in the use of Herbs," whose wife knew the bleeding techniques of humoral medicine, or when a woman near North Carolina's Sauratown Mountains returned a bundle of his lost specimens, recognizing their value.[56] On the whole, however, Lyon portrayed mountaineers as rustic people lost in a vast wilderness.

Lyon was not alone in making these critiques. Indeed, his assessments of southern mountain residents were quite tempered compared to some others. André Michaux described mountain homes as best characterized by "filth."[57] His son also recorded many such criticisms. Cumulatively, François declared mountaineers intemperate ("I do not believe that there are ten in a hundred who could resist the temptation of drinking as long as [the liquor] lasted"), inconstant (of an "unsettled disposition, which frequently, from the slightest motives, induces them to wander hundreds of miles in the hope of meeting with a more fertile soil"), and indolent. He declared that the primary occupation of western North Carolinians, so far as he could tell, was bear hunting.[58] Along the banks of the Ohio River in today's West Virginia, he castigated slash-and-burn agricultural techniques and shuddered at remembrance of the numberless fleas "with which most of the houses were filled." In summary, he claimed that nowhere else in his travels in the United States had he encountered such squalor.[59]

Few scientific travelers were harsher in their criticisms of southern Appalachian people than Elisha Mitchell, who visited western North Carolina in the late 1820s. Mitchell was a geologist by training and a minister by inclination, as well as a professor at the University of North Carolina at Chapel Hill. He was an avid observer of all facets of natural history during his trips to the mountains and paid close attention to regional plants. With the eye of a scientist he commented on wind-stunted Fraser firs on the highest Appalachian peaks, noted "a good many plants that were new to me," and took time out of traveling to dig "a root of ginseng."[60] Individual plants thriving in challenging environments may have been charming, but Mitchell found the landscapes shaped by people less endearing. Leaving the Piedmont for the mountains to the west of Wilkesboro, North Carolina, he noted that the land was "fertile but much less beautiful than that through which I passed yesterday. Something raw and countrified about it."[61] Deeper in the mountains, near Jefferson, the rough country and sparse homesteads sent Mitchell into rumination. In his journal he took the opportunity to record his belief in the power of nature over nurture, writing that "it is not in seclusion that the human mind receives its fullest development." He went on to claim that the mountain residents were interested only in "low and grovelling and brutal objects" rather than noble pursuits.[62] Later, whether from accumulating disgust or a love of moralizing, Mitchell concluded that remote mountain terrain broke down the most fundamental of civilized relations: "In a rude hunter's state of society, the women become schquaws [*sic*], very pretty ones, but schquaws notwithstanding."[63]

Thomas Nuttall too found much to critique along the Virginia and Ohio border. He wrote of local people "given to rambling about instead of attending to their farms, [and who] are poor and uncomfortable in every respect; but few of them possess the land on which they live." He largely attributed these conditions to environmental causes, arguing that they stemmed from the pursuit of "comfort and independence which they in vain seek in the solitudes of an unhealthy wilderness."[64] In southwestern Pennsylvania Nuttall also turned up his nose at residents who he believed were "chiefly Irish . . . indigent, and considerably deficient in prudence and cleanliness."[65] Likewise, Rafinesque felt quite adventurous when journeying through the districts of eastern Kentucky, wondering how people could live in those "wild and hilly places, nearly unsettled, having sometimes to go 14 miles without meeting a Cottage."[66]

Gray, though remarking favorably on the kindness of western North Carolinians, also sighed at their lack of "general intelligence." He wrote that they "could scarcely be made to comprehend the object of our visit."[67] Gray's wife recalled that he found southern mountaineers "woefully ignorant, especially of everything relating to Christianity." From his northeastern home, Gray corresponded with Episcopal Church officials and successfully lobbied for a mission to be established in Valle Crucis, North Carolina, for the purpose of improving the region's educational opportunities and morality. On his return to the mountains years later, he lamented the lack of moral progress, complaining of seeing a Bible, which had been distributed by the mission, being used as a child's scrapbook.[68] Collectively, botanizers at best found southern mountaineers hale and hearty but charmingly ignorant. At worst they saw them as the dregs of society, their intellect and morals eroded by the savage nature of the mountain wilds.

An important question here is whether these botanists' assessments of Appalachian nature and people reached a broader public. As private observations they are intriguing but mean little. In many instances, however, collectors' descriptions did find a larger audience. For example, William Bartram published his massive journal, which included his mountain travels, with a Philadelphia printer in 1791, and it was reprinted in London the following year. It would become a classic of American science and literature.[69] The elder Michaux tended to publish more technical plant guides, such as his definitive study *American Oaks*, and in those manuals he stressed the remoteness of some Appalachian specimens, such as the chinquapin oak.[70] Others, close to Michaux, drew on his unpublished journal shortly after his death to describe Appalachia for European and American audiences, dwelling at length in magazine articles on his accounts of dangerous terrain and isolated people.[71] François André Michaux also drew on his father's account, including references to his comments, when shaping his own tour of the region and published his descriptions of the mountains in book form in London in 1805, as *Travels to the Westward of the Allegany Mountains*.[72] Rafinesque published a scientific journal—often containing his own articles—that frequently described Appalachian landscapes as well as the region's natural history. And Gray often relied on his collecting trips to provide fodder for both scientific papers, on especially interesting specimens, and more general articles, describing his travels and regional observations, published in popular magazines.[73] In sum, not all botanizers' thoughts reached public eyes, but a significant portion did find their way into publication on both sides of

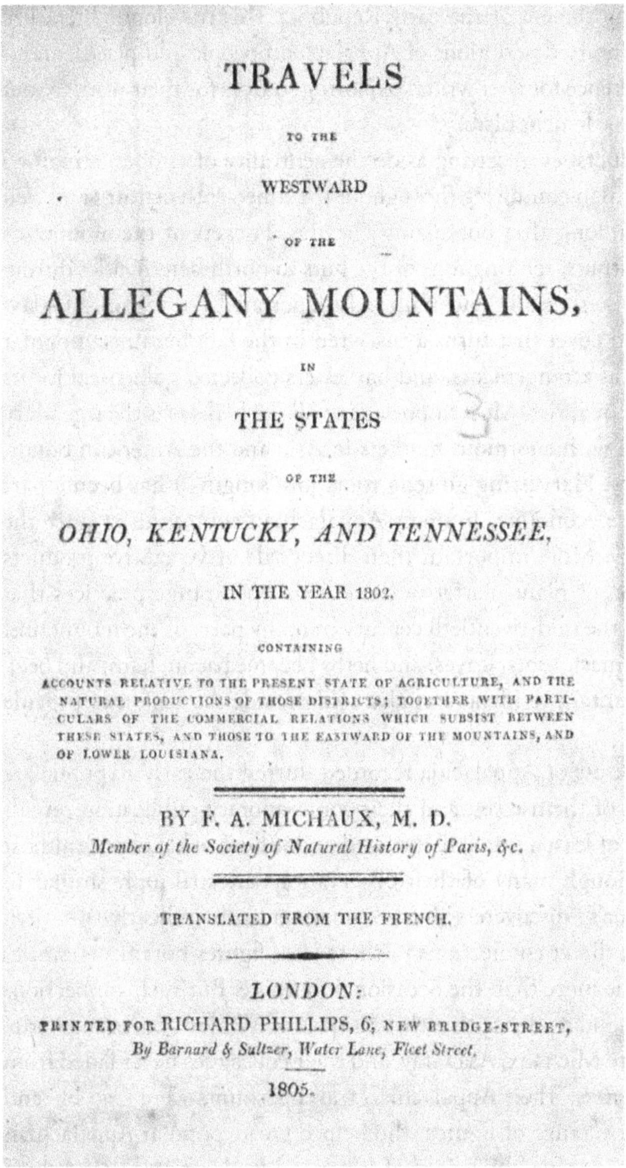

TRAVELS

TO THE

WESTWARD

OF THE

ALLEGANY MOUNTAINS,

IN

THE STATES

OF THE

OHIO, KENTUCKY, AND TENNESSEE,

IN THE YEAR 1802.

CONTAINING

ACCOUNTS RELATIVE TO THE PRESENT STATE OF AGRICULTURE, AND THE
NATURAL PRODUCTIONS OF THOSE DISTRICTS; TOGETHER WITH PARTI-
CULARS OF THE COMMERCIAL RELATIONS WHICH SUBSIST BETWEEN
THESE STATES, AND THOSE TO THE EASTWARD OF THE MOUNTAINS, AND
OF LOWER LOUISIANA.

BY F. A. MICHAUX, M. D.

Member of the Society of Natural History of Paris, &c.

TRANSLATED FROM THE FRENCH.

LONDON:

PRINTED FOR RICHARD PHILLIPS, 6, NEW BRIDGE-STREET,
By Barnard & Sultzer, Water Lane, Fleet Street,

1805.

FIGURE 2.2. Title page of François André Michaux's *Travels to the Westward of the Allegany Mountains* (1805).

the Atlantic during the era of the early Republic. This travelogue literature provided contemporary descriptions of Appalachian people and places, and it would serve as reference for later writers exploring sources for their works about the region's early settlement history.

Botanical products, even setting aside the centrality of timber, remained critical in Appalachian commerce throughout the nineteenth century and well into the twentieth, long after botanizing's heyday. Pockets of the mountains specialized in chestnuts, sending tons of the nuts to northeastern cities during the holiday season, especially once railroads penetrated the region.[74] Galax, a perennial ground cover that turns a rusty red in the fall, became a popular component of floral arrangements, and harvesters collected goldenseal for its reputed medicinal qualities. Most important of all the herbs was the legendary ginseng, a heal-all with enormous markets in Asia and the American botanical drugs industry. Harvesting ginseng roots (or "sangin'") has been a part of local subsistence economies in many Appalachian communities since the eighteenth century. More important than direct sale of vegetative products was the conversion of plant matter into meat via open range practices that persisted well into the mid-twentieth century in many parts of the mountains. Countless tons of mast, roots, leaves, and herbs became bacon, ham, and beef, consumed by mountain residents but also sold outside the region to provide some income.[75]

Botanists' accounts of Appalachia recorded during the early Republic are interesting in and of themselves, and their contemporary publication reveals that some reached at least a small, if scattered and influential, audience almost immediately. Although many of their observations are strikingly similar to those of Appalachia's "discoverers" in the late nineteenth and early twentieth centuries, without direct connections to those later figures botanizers' stories might warrant little more than the occasional footnote. But such connections do exist. There is solid evidence that the critiques of William Bartram, André and François André Michaux, Asa Gray, and their colleagues never faded from popular consciousness. Their Appalachian travel accounts were read by, and influenced, a diverse range of figures who helped create popular Appalachian stereotypes a century after the height of botanical exploration. They helped imagine and define "a strange land and peculiar people" long before Will Wallace Harney coined the phrase in 1873.[76]

One group of late-nineteenth-century scholars interested in these travel accounts was a new generation of botanists. As the field moved away from being an avocation for wealthy amateurs and collectors for hire and became a profession, complete with university positions and graduate schools, botanists became increasingly interested in recording and defining their discipline's history. The American most influential in these efforts was Charles Sprague Sargent, director of Harvard's Arnold Arboretum and in his day the nation's most important botanist. Sargent had a deep interest in the prominent botanists of decades past and made a secondary career of sorts editing, annotating, and publishing their travel narratives and correspondence. In the 1880s Sargent introduced Americans to André Michaux's travel diary (in the original French) through an annotated journal article as well as in book form.[77] Sargent would do a similar service for the previously unpublished papers of Asa Gray, assembling and editing two volumes that appeared in 1889.[78]

Although Sargent's work (and to a lesser extent that of other botanists) kept Michaux and Gray in the scientific eye, other figures did more to popularize the travels and opinions of botanizers. For example, Reuben Gold Thwaites's work as an editor helped ensure that certain botanists would become primary sources for professional historians of the early twentieth century. Between 1904 and 1907 Thwaites published thirty-two volumes of edited American travel narratives, some of which dealt with scientific collectors in Appalachia. Like his famous contemporary Frederick Jackson Turner, Thwaites was a historian of the West who viewed Appalachian settlement as an early expression of America's inevitable westward expansion, its "manifest destiny." One volume comprised an English translation of André Michaux's journal, and he reprinted the younger Michaux's *Travels* as another. Thwaites also reproduced Nuttall's book on the Arkansas Territory, which included a section on his passage through Appalachia.[79] Subsequent historians, such as Samuel Cole Williams, would include these works by botanical collectors in their accounts of early Euro-American culture in the southern mountains.[80]

Prominent figures in the literary "discovery" of Appalachia also drew on the accounts of botanists from a century earlier when describing the region's longstanding wildness. Two books published in 1913 illustrate the lasting influence of botanizers' descriptions of places and people. Biologist and nature writer Margaret Morley's *The Carolina Mountains* introduced plants and environments first described by William Bartram, André Michaux, and Asa Gray to a general audience. When describing the peaks and people of western North Carolina she mentioned Mitchell's scientific work and death on his namesake mountain, referenced the fame of Roan Mountain with early botanists, and included the story from Michaux's journal of his Grandfather Mountain ascent and his mistaken belief that the peak was the highest point in North America.[81] The same year, outdoor writer Horace Kephart's *Our Southern Highlanders*, one of the most popular early-twentieth-century accounts of Appalachia, used the old botanists' narratives to similar effect. He wrote of the formative environment of the southern mountains: "It was the botanist who

" . . . Powerful steep and Laurely . . . "

FIGURE 2.3. Illustration from Horace Kephart's *Our Southern Highlanders* (New York: Outing Publishing, 1913), portraying a remote mountain environment that Kephart believed little changed from a century earlier.

discovered this Eden for us." And he described some of the rare specimens cataloged by William Bartram as well as the elder Michaux's impressions of the region (also incorporating the now famous story of Michaux on Grandfather). Kephart carried his narrative of scientific discovery through the work of Asa Gray and his descriptions of mountain nature. In the remainder of *Our Southern Highlanders* Kephart used these descriptions of Appalachian environments to argue for the environmental roots of what he saw as mountaineers' cultural backwardness. In this way his explanations duplicated the claims of the botanizers from a century earlier.[82]

Regional fiction writers had a harder time weaving botanists into their narratives, at least by name, but their Appalachian backgrounds and characters echoed the old themes. Novelists such as Mary Noailles Murfree and John Fox Jr. depicted rugged, wild mountain environments that contributed to equally rustic characters. Western North Carolina novelist Shepherd Dugger managed, through a tortuous plot twist, to include some real botanizing history in his most famous work, *The Balsam Groves of the Grandfather Mountain*, which was first published in 1892. Dugger's character Mr. Clippersteel, when leading a party hiking on Grandfather Mountain, recounted the story of André Michaux's labors on the same slopes "to collect seeds, shrubs and trees for the royal gardens." Clippersteel then produced a copy of Michaux's journal—perhaps Sargent's annotated edition?—and began to read word for word from Michaux's journey up the mountain, thus weaving the real botanist's account of mountain nature into the fictional narrative.[83] Dugger explained away Michaux's mistaken elevation calculations, arguing that the power of Appalachian "environments upon the aesthetic mind" shaped his perceptions of the region as surely as they shaped local people and culture.[84] The novel even included an excerpt of Sargent's introduction to Michaux's annotated journal as an appendix of sorts, as well as fragments of Michaux's 1794 journal as translated by Dugger.[85] A later edition of *The Balsam Groves* would add a brief history of Elisha Mitchell.[86] Through this pastiche, Dugger at once affirmed the botanists' perceptions of the historic region and testified to its continued validity.

That the cultural discovery of Appalachia spanning the turn of the twentieth century drew on botanists' descriptions of the earlier region has gone all but unnoticed. Appalachian historians, such as David Hsiung and Wilma Dunaway, have convincingly established the region's connectedness to the broader world from its initial Euro-American settlement.[87] Although there is some truth in portrayals of the mountains as relatively isolated, the histories of botanical collectors do reinforce arguments for Appalachia connectivity, highlighting economic and intellectual ties between the southern mountains and the broader world. And botanists' notes, journals, and publications demonstrate one early expression of describing Appalachian exceptionalism. This exceptionalism consistently linked rugged nature to isolation and cultural difference, an interpretation driven by the extraction of natural resources in the form of plant specimens for profit and prestige. The important point here is that Appalachian exceptionalism developed not in opposition to the realities of interconnection but out of them. The reliance of later cultural critics of the region on the writings of these botanists, influenced by and repeating their observations for a new generation of Americans, demonstrates the lasting power and influence of these early accounts.

These connections highlight the root of the environmental determinism that became instrumental in the idea of "Our Contemporary Ancestors," immortalized by Berea College president and progressive social activist William Goodell Frost in 1899. In describing Appalachia as a problem area in search of reform, Frost invoked ideas of the frontier, much like Frederick Jackson Turner did in his classic treatise on American identity, imagining that something about Appalachian nature created a distinct culture and people. In contrast to Turner, Frost and other "discoverers" of Appalachia believed nature did more to freeze development than to forge a new American identity, more to create distinctiveness than to homogenize diverse cultures. If Turner's frontier made Americans by stripping away European habits and replacing them with egalitarian individualism and the habits of democracy, Frost's forgotten frontier preserved English and Scots-Irish traits in a way that was at once interesting and queer.[88]

It would be easy to take the comparison too far, but there is much in the written accounts of collectors that Frost must have found familiar, an association of a distinct place and a distinct people, with each side of the equation reinforcing the other. Here, believed the botanists, was nature stretching her green hand over culture, shaping lives and customs as surely as climates, forests, and soils; certain people as well as plants might have environmental niches. All the elements that would become part of an externally imposed Appalachian identity were there: poverty, isolation, the power of geography, ignorance, despair, and beauty, all uncovered, described, and reified by the external, professional gaze, with an economic and intellectual interest in exoticism. Botanists' language concerning both mountain environments and mountain residents evoked—and helped invoke—the discovery of Appalachia a century later, to such an extent that an attuned ear might hear echoes of the botanizers in the work of John Fox Jr.

CHAPTER 3

Gold

The Rise, Fall, and Rebirth
of Southern Gold Mining

SUTTER'S MILL, in the foothills of California's Sierra Nevada, holds
a storied place in the American imagination as the birthplace of the nation's
obsession with gold. In the hills and mountains surrounding the mill, the gold
rush that followed the 1848 western strikes made California the nation's golden
heartland. But California's wealth and its romanticized gold rush days obscure
a longer history of American gold mining, one with deep roots in the southern
Appalachian Mountains. A history of the southern mountains (and especially
of North Georgia) from the 1820s until the Civil War is a story of America's
first major gold rush, a southern one that preceded the California frenzy. This
antebellum Appalachian gold rush served as one of the many mountain and
southern paths to industrialism; what began as a small extractive enterprise
quickly became complex and capital-intensive. Although this Appalachian
gold rush figures prominently in Georgia history and tourism, it has not received its warranted degree of broader scholarly attention. It was a crucial de-
velopmental period in Appalachian industrial history and in the nation's mineral economy, as gold became one more commodity connecting the mountains
to the lowlands. Perhaps of even greater significance, the gold rush reinforced
ideas about the appropriate use of Appalachian nature as a marketable commodity, repeating commercial patterns present in the fur and plant trades.

This is a story of destruction as well as creation; in this the gold rush was
more akin to the fur trade in its ruthless environmental exploitation than to
botanical collecting. Ideas about the best and most profitable use of nature had
real consequences on the ground. Appalachian gold fever remade local economies, societies, and environments. In pursuit of wealth, miners ripped apart
streambeds and hillsides, cut down forests, and erected miles of wooden flumes
and towns of flimsy shacks, and stamping mills crushed ore to dust in an effort
to separate gold. Public and private mints sprang up to transform precious
metal into currency, and, with the help of the state and federal governments,
speculators obsessed with the prospect of riches helped drive the surviving

Cherokee people from Georgia on the "Trail of Tears." In its all-encompassing drive to extract minerals from the earth with little regard for the environmental or human costs, gold mining looks quite similar to the regional coal mining that followed it a half-century later.

The Georgia gold rush affected other important aspects of the American political economy as well. A few historians have made strong arguments that this early gold fever served as a model for the better-known gold rushes that followed it in California, Colorado, the Dakotas, and eventually the Klondike.[1] Southern gold also provided much-needed precious metal for mints attempting to create fiscal stability and confidence in currency in a time when both were sorely lacking. According to historian Stephen Mihm, this was an era when the United States was a "nation of counterfeiters."[2] Americans suffered from general uncertainty about the foundations of the nation's wealth—faith in credit and the value of land had wavered in the face of burst real estate bubbles like the one accompanying the Yazoo lands speculation—and Georgia gold offered hope of tangible wealth to shore up this lack of consumer confidence.[3]

Gold mining also proved important for Georgia's economy, although its benefits were temporary. Agriculture was far and away the state's most important source of income in the nineteenth century, and cotton was entrenched as its chief commodity. By the nation's first agricultural census in 1840, Georgians were producing more than 163 million pounds of cotton each year.[4] For all of cotton's preeminence and perhaps because Georgians were so dependent on the staple, critics frequently stressed the need to diversify the state's economy.[5] Gold mining provided the opportunity to do so in North Georgia (where growing cotton was impractical), and throughout southern Appalachia miners had dug roughly $40 million worth of gold by the start of the Civil War, most of it in Georgia. Gold would never supplant cotton in Georgia's economy, but mining's economic impact on the state and nation was certainly significant.

The historiography that does exist on the Georgia rush—some of it quite good—focuses on its beginnings and pays special attention to the importance of gold discovery for the Cherokee Removal.[6] Although the early years of Appalachian mining were historically significant, the decades that followed were consequential too. Largely overlooked, the years immediately preceding the Civil War highlighted the remaining connections between the southern Appalachian mineral economy and distant mining districts and financial centers like New York, even after California took center stage. The 1849 exodus of miners to the West Coast stripped Georgia of skilled labor, but new mining techniques promised to lure at least some of them back. As the Civil War loomed, engineers with experience in the West and property owners in North Georgia applied California techniques of hydraulic mining—costly and elaborate processes using canals, flumes, and high-pressure hoses to wash away huge quantities of mountain soil to expose hitherto hidden gold deposits—with the backing of northern capital. Like so many mid-nineteenth-century Americans falling under the spell of science and new technology, mine owners and speculators imagined that technical expertise and ambitious investment could solve their troubles. New techniques, they speculated, would provide a more complete mastery of nature and once again make Georgia a center of national industry, thus reversing the flow of energy, wealth, and population from West to East.

The hydraulic mining underway by the start of the Civil War recaptured the frenetic speculative spirit of the early 1830s, creating a brief second gold rush in Appalachia. Georgia miners undertook the renewed extraction of natural resources on an industrial scale, efforts that drew parts of the southern mountains—at least temporarily—fully into the national economy in ways that presaged the timber and coal booms of the postbellum period. Indeed, by 1861 the industry's reliance on professional engineers, distant shareholders and speculators, and expensive extractive technologies very much resembled the structures and techniques of the large coal companies that would come to dominate Appalachia a half-century later. It is too simple to say that Appalachian gold mining led directly to late-nineteenth-century coal mining, but it is clear that the ideas about nature and the efficient systems for exploiting environmental resources that dominated coal mining were present in North Georgia before the Civil War. Gold mining would gradually cease to be an important economic force in the southern mountains (hydraulic mining would ultimately fail to resurrect the dynamism of the early Georgia gold rush), but miners and ideas about the industrial uses of the environment were there to stay.

ᴄᴐ

Like many events that have become part of the American historical narrative, the Georgia gold rush has its origin myth, a tale that provides "a natural justification and clarity so that interpretations become statements of fact."[7] This particular myth begins in the fall of 1828, with an Appalachian farmer named Benjamin Parks, who was hunting or perhaps returning from moving his cattle to a salt lick along the Chestatee River or both (accounts vary). Strolling through the hardwoods he stumbled over a rock, bent over to look at the offending stone, and noticed that it was flecked with gold. Despite Parks's efforts to hide the discovery, word quickly spread, and soon thereafter the gold rush was on.[8]

The Parks story has not gone unchallenged. A range of other claims to and about the first discovery exist, alternately crediting the find to various other local white residents, a slave, a Cherokee hunter, and a pair of English miners. Some agree that Parks was first to find gold but that the discovery occurred in a different, unspecified location. Even the year 1828 is in doubt. David Williams, a historian who has devoted a great deal of study to the origins of the rush, questions it, saying "no firm evidence for gold in Georgia is found until 1 August 1829."[9] The community of Villa Rica (in Carroll County, southwest of the Parks strike) also claims to have been the first to unearth gold in the state, dating the local discovery to 1826.[10]

Whatever the truth of stories about the strike, the spread of mining in Georgia is perhaps best understood as a logical extension of smaller prospecting efforts in the Appalachian foothills to the northeast. As early as the 1790s, farmers uncovered the occasional gold nugget in western North Carolina, including the largest specimen ever discovered in the United States, a twenty-eight-pound lump found in Cabarrus County in 1799.[11] Southern miners had made a series of strikes from Maryland through North Carolina during the 1820s. In North Carolina in particular, substantial mining operations developed, and the geologist Denison Olmsted circulated a description of regional mining to a national audience as early as 1825.[12] By the late 1820s, the rush in the Carolina goldfields was at such a frenzy that University of North Carolina geology professor Elisha Mitchell cursed the hucksters who preyed on people's gullibility. In his diary, he derided the divining rod charlatans—"a race of vermin who infest this country and share the confidence of the people"—who were swarming the region and fleecing hopeful miners.[13] By the end of the decade, North Carolina miners had turned their gaze to the southwest, looking for regions with similar geology or following productive watersheds downstream into South Carolina, Georgia, and Alabama.[14]

Whatever the true origin of the Georgia rush, prospectors flooded North Georgia (most of it still officially Cherokee territory), and within a few months "the first genuine gold rush in American history was going full tilt." By 1830 there were as many as ten thousand prospectors in the state, with perhaps four thousand of them lining the banks of the Yahoola River (in present-day Lumpkin County and today known as Yahoola Creek).[15] Likewise, observer Yelverton King claimed that there were "between four and five thousand men engaged in digging and searching for gold in that part of the [Cherokee] nation [land] attached by the laws to the county of Hall."[16] It was a population "constantly increasing with a rapidity almost too incredible to relate."[17]

Georgia officials moved quickly to secure the interests of American miners. In 1830 the state legislature declared its authority over all Cherokee territory within Georgia's boundaries, and Congress passed the Indian Removal Act, which promised to force all Native Americans from the state. State officials

FIGURE 3.1. A modest North Carolina gold rush predated strikes in Georgia and encouraged the idea that the southern mountains were rich in mineral deposits. "View of the Gold Hill Works," 1857. *Harper's New Monthly Magazine.*

then began auctioning off Indian land; white Georgians could claim the confiscated territory in either 160-acre agricultural lots or 40-acre mining tracts within the gold belt. President Andrew Jackson gave these state actions tacit authority by withdrawing federal troops from the gold district. In 1838 the land grab culminated in the tragic forced march of Georgia's remaining Cherokee people to Indian Territory, in present-day Oklahoma.[18] The initial gold lot auctions kindled an incredible speculative fever. A single county's land lottery drew twelve thousand interested Georgians, and a total of 133,000 state residents registered for the thirty-five thousand available gold lots. Investors hovered around the margins of the drawing-day crowds as the wooden lottery drums spun and officials called out the names of winners one by one, waiting to make offers to the lucky citizens who won the most promising lots. In the midst of the 1833 auction frenzy, the *Savannah Georgian* reported that "fame [had] put the lot 1052-12-1 at the high valuation of $100,000."[19] Land changed hands almost too quickly for courts to follow, and the region was rife with hucksters, confidence men, and downright frauds intent on claiming their own slice of the profits generated by gold fever.[20]

As land claims were secured (and in many cases before such legal niceties), prospectors went to work. Later critics declared the early mining in North Georgia primitive and inefficient, with miners scratching the surface for obvious gold rather than systematically processing the soil. As was the case with the fur trade, these portrayals equated this mining with the frontier, a hardscrabble existence that would fade away as civilization developed and disseminated better technologies and techniques.[21] There was some truth to these claims. The first miners sought accessible surface gold that had eroded from veins of underground gold ore, or "lodes." They could find this "placer" gold with the most basic equipment: pans (often made from a skillet), balances, rockers, hoes, picks, and shovels. A prospector engaged in basic placer mining identified a likely looking stream-side site based on his experience and understanding of soils and hydrology, dug a bit of alluvial dirt and gravel with hand tools, and placed it in his pan. Adding water to the pan and swirling the mixture with a practiced hand slowly separated the lighter dirt and gravel from the heavier gold particles. A practiced panner cast waste out of the pan with each swirl, leaving gold dust behind. Rockers, wooden boxes on curved frames that could be rocked back and forth (the crudest were simply hollowed-out logs), functioned much the same way. The rocking action separated stone and other debris from gold-bearing soils, and the latter were trapped in sieves or wooden ridges on the rocker's bottom. The resulting gold dust was weighed on a balance and then often carried in the most primitive of holders, the hollow quill of a goose or turkey feather.[22] The era of gold accessible via basic panning was short, however, as miners quickly worked over the best placer material. By 1836, prognosticators were already declaring that the heyday of panning had passed and that the Georgia goldfields were in decline, but more elaborate techniques and technologies soon found their way to the goldfields.[23]

Placer mining was quickly replaced by more complicated vein mining that followed gold-bearing lodes into the ground. Although later miners would also criticize these efforts as primitive, limited by "the unskillfulness of those who worked, not being accustomed to underground operations," they represented a real intensification of mining efforts.[24] Most of these cuts were shallow. Miners dug into a hillside only as far as the overhanging earth would support itself, resulting in a mine that looked much like a shallow cave. On occasion, however, the mine works evolved into true tunnels, with timber supports to prevent ceiling falls and pumps to keep the works dry.[25] Large scale earth-moving also took place through the techniques of alluvial washing, a more elaborate form of basic placer mining. Miners located a promising site in or near a stream, dug down through the gravel layer (or alluvium), and shoveled the material yard by yard through rockers and sluices, keeping the gold dust that emerged and sending the rest of the earth and stone downstream.[26] Just fifteen years after the start of the rush, some prime alluvial deposits had been worked three or four separate times, and miners' labor had begun to reshape the mountain

BORING.

FIGURE 3.2. Most antebellum Appalachian gold mining employed picks, pans, and rockers, but rich veins encouraged rudimentary tunnel mining, as depicted in this illustration of a gold mine shaft in North Carolina. "Boring," 1857. *Harper's New Monthly Magazine.*

environment. As mine supervisor Thomas Clemson noted, these were hardly casual endeavors: "The amount of labor that has been expended in this region is prodigious. The streams have been turned from their courses, and the banks and alluvion [*sic*] have been dug down to the slate."[27]

Within a few years of the discovery of gold, Georgians pioneered the development of mining equipment that soon spread across the nation. Historian Otis Young Jr. credits these innovations to the particularities of North Georgia geology. The relative lack of nuggets in local gold deposits, and the region's frequent intermixture of clay and fine gold dust, presented serious mining challenges that could only be overcome by evolving technology. In response, inventors introduced numerous machines to surmount these challenges or to tap the wealth of gold thought to be located in the rivers. The continuous gold washer used a diverted stream's flow to wash soil through a long trough, where wooden or metal riffles trapped gold particles. The double compartment rocker more efficiently separated gravel from the heavy silt that might prove to be "pay dirt." Small dredge boats raised potentially rich sediment from the bottom of Georgia rivers. Simple diving bells accomplished a similar purpose, trapping air that a miner could breathe to work underwater. And substantial water-powered crushing mills used cast iron strikers (called stamps) to pound quartz ore and release the gold dust adhering to the rocks.[28]

The transition to vein mining and alluvial washing increased the business's complexity, requiring more detailed bookkeeping, expensive machinery, and a plethora of tools. For example, one 1855 account book recorded a typical order of tools shipped from Brooklyn to outfit a stamp mill for the Hart County Mining Company. It included a steam engine to power the mill and "an Anvil, Screwplate, Bellows, Monkey Wrench, Carpenters Level, Hand drill, Hammer, Grindstone, Machinery oil, Pair of double Blocks, & 50 fathom Rope Belting." The cast iron stamps alone were quite an investment, at nearly $1,300 for eight.[29]

Mining tools and machinery were expensive, and so from an early date various forms of capitalization were critical to success, further adding to the complexity of mining. At the most basic level, many landowners supported mining operations through surplus labor and capital accumulated on plantations, connecting local industrial activity to the broader southern agricultural economy. Most crucial was slave labor, which appeared early and often in the gold district as planters from the Piedmont and Lowcountry looking to diversify their income turned to mining during the winter, when cotton work was slack and the lack of summer vegetation made mining easier. Leasing slaves to mining operations was also popular. In this fashion, Georgia gold mining served as a supplement to the state's cotton plantations rather than competing with them. Forty-acre Appalachian mining tracts became a different sort of "back forty" for some lowland plantations.[30] The same held true in the North Carolina gold district, where in 1833 in Burke County as many as five thousand enslaved African Americans labored at gold mining.[31] (A few free blacks also labored in Georgia's mines, and at least one black man, James Boisclair, purchased his own plot of land despite a state law prohibiting the sale of gold lots to African Americans.)[32]

Planter investment was important, but capital from more distant sources quickly dominated regional mining operations, as foreign and northern investors established companies and camps to make Appalachian gold extraction more efficient and profitable. Where gold flowed out, money flowed in. A British company managed a North Carolina camp worked by "Germans, Swiss, Swedes, Spaniards, English (this almost invariably means Cornish), Welsh, and Scots. Thirteen languages were said to be represented."[33] During the 1850s, North Carolina's northern-owned Gold Hill works employed three hundred men, mainly Cornish and African American miners, and used steam engines to power most of its machinery.[34] As early as the 1830s, agents like New York City's James F. D. Oldenberg specialized in bringing northern investors and Georgia gold-lot owners together, and a number of companies emerged that were capitalized with half a million dollars or more.[35] By 1840, incorporated businesses like the Augusta Mining Company, Naucoochy Mining Company, Pigeon Roost Mining Company, Cherokee Mining Company, and Lumpkin County Mining and Manufacturing Company dominated mining efforts in North Georgia.[36]

As was the case with deerskins and botanical specimens, parts of Appalachian nature entered broader commercial exchanges as gold left the mountains. Increased gold mining stimulated the national economy, which had been starved for hard currency prior to the southern gold discoveries.[37] At first the gold flowed to the federal mint in Philadelphia, but as production increased, local minting of coins became increasingly attractive. By the late 1830s, the federal government established branches of the national mint in Charlotte, North Carolina, and Dahlonega, Georgia. In its first year of operation alone, the Dahlonega mint stamped more than $100,000 worth of gold coins, and it eventually coined $6.1 million.[38] Two private mints, one in Gainesville, Georgia, and the other in Rutherford County, North Carolina, also produced gold coins that served as legal tender throughout the Southeast. These private ventures were no small affairs. Otis Young Jr. asserts that the Rutherford operation was "the most productive private mint in national history," ultimately issuing more than $3.6 million in coins and ingots.[39]

Big endeavors attracted big names, and no figure in the early Georgia goldfields loomed larger than John C. Calhoun. South Carolina's fiery voice supporting the institution of slavery and states' rights saw opportunities in gold, perhaps because exploitation of southern lodes promised greater regional autonomy. Calhoun purchased a mining tract on the Chestatee River in Lumpkin County and employed his son-in-law and mining engineer, Thomas Clemson (later benefactor of Clemson University), to manage the business. Some accounts contend that the Calhoun tract was the same plot of land where Benjamin Parks made the discovery that started the rush.[40] The Calhoun mine proved one the region's most productive, employing a substantial

FIGURE 3.3. John C. Calhoun, who owned one of Georgia's largest early mines. Alexander Hay Ritchie, engraver, from a painting by Thomas Hicks, from a daguerreotype by Matthew Brady, circa 1852. Courtesy of the Library of Congress.

slave labor force to dig cuts and tunnels into hillside veins. The extracted ore was then crushed in a stamp mill, where advanced techniques helped capture gold dust. Among these technological innovations was the use of mill plates covered with liquid mercury. When crushed ore passed over the plates, gold particles chemically bonded to the mercury, and the precious metal was then separated from the amalgam through distillation, which vaporized the mercury.[41] Calhoun was involved in other goldfield endeavors too, encouraging the development and growth of a boomtown: Auraria, Georgia.[42]

Whatever people thought of the relative sophistication of gold operations, they consistently acknowledged their steep environmental costs. According to a correspondent to *Merchants' Magazine*, "on approaching Dahlonega I noticed that the water-coarses [sic] had all been mutilated with the spade and pickaxe, and that their waters were of a deep yellow; and having explored the country since then, I find that such is the condition of all streams within a circuit of many miles. Large brooks (and even an occasional river) have been turned into a new channel, and thereby deprived of their original beauty. And of all the hills in the vicinity of Dahlonega which I have visited, I have not yet seen one which is not actually riddled with shafts and tunnels."[43] A mining engineer, reflecting in the 1850s on the physical effects of the early gold fever, was troubled. Gold production was a boon for the nation, he asserted, yet in reflective moments he could not help but imagine that "the poor Indians saw with dismay their beautiful hill-sides, where they and their fathers had chased the deer for centuries, occupied by these centres of vice and immorality, and their lovely valleys and cool dells dug over and rendered hideous to the sight."[44] Another journalist was more succinct: "The whole population is engaged in digging for gold; and the face of the country for many miles presents the appearance of new made graves."[45]

Nature, then, could be corrupted, but it also corrupted gold seekers, at least according to outside observers. These critics believed that the primitive labor of extracting minerals from the earth, close contact with mountain forests and streams, and perhaps something in the very nature of gold itself turned civilized men and women wild.[46] Maybe, they speculated, this wildness was caused by the local environment, which was a wooded tangle where, "were it not for the hidden treasures of the mine, no human habitation would have intruded on the haunt of the wild-deer and wolf."[47] Thomas Clemson's wife, Anna (accompanying her husband to his work supervising Calhoun's mine), described the goldfields as the most primitive of places, with roads "but a path cut through the woods," snaking here and there by homes as poor as "the roughest negro house."[48] It was a region where the people walked with a stoop, staring at the ground, where "it is common after a rain to see the inhabitants busily engaged picking up gold in the streets."[49] Thanks to this frontier image, the gold district became a popular destination for religious missionaries from the Northeast, with some reaching the area as early as 1833. H. P. Pitchford, arriving on the Chestatee River in 1835 to spread God's word, wrote that "a great many people here are as busy as if the mines were the Church, the gold religion, and the tippling shop a little heaven."[50]

These narratives of gold as corrupting and its extraction as demeaning continued well past the early years of Georgia mining. The same year that the California gold rush lured ambitious people west by the thousands, a correspondent for *Knickerbocker* magazine described Appalachian miners as unkempt, ignorant, and uncivilized. Even a man with obvious pretensions to southern planter aristocracy—he was "worth forty negroes"—"lived in a log-house and could neither read nor write."[51] And in Dahlonega, scene of frenzied economic opportunity, life remained rough, lived out in unpainted houses and saloons. Gold might corrupt body and soul. According to moralizers, "The very wealth of the country has caused the ruin of many individuals."[52]

Southern writer William Gilmore Simms codified these connections between rough nature and rough people in his 1834 novel of the gold district, *Guy Rivers: A Tale of Georgia*. Simms, like his better-known northern contemporary James Fenimore Cooper, believed in the power of the American frontier to shape culture. Although he likely never visited the goldfields in person, from the first pages of *Guy Rivers* he invokes an atmosphere of environmental determinism. Simms opens the novel with a memorable excoriation of the North Georgia landscape: "If not absolutely desolate, [it] has, at least, a dreary and melancholy expression, which can not fail to elicit, in the bosom of the most indifferent spectator, a feeling of gravity and even gloom." It was a countryside of "generally steril [*sic*] character" that seemed "utterly deficient in resources"—except, of course, for gold.[53] And although nature seemed to promise through this precious metal to redeem the mountains "from the curse of barrenness," life in the goldfields ultimately proved degrading and dehumanizing. A Romantic in tendencies, Simms frequently shifted to brutal realism in his portrayals of life on the gold frontier, with Appalachian nature anthropomorphized as a malevolent actor.[54]

Despite these marginal worries, during the 1830s and 1840s the North Georgia goldfields served as America's golden mecca, a place for those who dreamed of mineral riches. Otis Young Jr. calculates that in total, southern mining produced approximately $40 million worth of gold before 1861 (although he admits this figure is highly speculative). About $9 million of the total came from North Carolina mines (approximately 450,000 ounces of gold), and Georgia miners produced most of the balance. This figure pales in comparison to the estimated $550 to $680 million dollars worth of gold mined in California by 1861, but from the 1820s until 1848 the Appalachian South provided virtually all of the gold mined in the United States. Southern gold was thus a critical economic resource for a specie-starved nation.

Although the first few years of the rush have drawn the bulk of the historical attention, the late 1830s and 1840s were the most productive era of Georgia mining.[55] Notwithstanding this continued productivity, Georgia mines lost their luster as quickly as they had risen to prominence when word of a new strike reached the East from California. Once "the wealthiest gold region in the United States," Georgia became just another launching point for journeys west.[56] Despite efforts of prominent locals to keep miners at home (such as the boosterism of Dahlonega mint assayer Matthew Stephenson, who, pointing to the ridge line looming over the town, famously shouted, "Why go to California? In that ridge lies more gold than man ever dreamt of. There's millions in it"), prospectors left in an exodus for the Pacific Coast in 1849.[57] Perhaps as many as five thousand Georgia miners departed for California in this initial rush, when even Calhoun's mine closed for lack of labor and an uncertainty in the further profitability of Appalachian mines.[58]

These Georgia emigrants would prove important figures in western gold districts throughout the 1850s, bringing muscle, mechanical inventions, and practical experience to the new mines, transferring lessons learned in Appalachian forests to the western mountains. Historian David Williams argues that "it was in fact the gold miners of the South who formed the core of the California 'forty-niners' and provided the expertise and technological know-how gained during a quarter century of mining Southern gold."[59] Georgians made their presence felt in the Colorado gold rush a few years later as well, when a group of miners just arrived from Auraria, Georgia, circulated some of the first reports of a Front Range strike in 1858. They named their new mining camp after their old hometown, and in 1860 the Colorado Auraria merged with a sister camp to form a new town: Denver.[60]

Western gold strikes did not spell the end of Georgia mining, however, and work in the southern mines during the 1850s is the least explored part of the Appalachian gold rush. Despite their relative obscurity, these years reveal much about the intensive, national connections of this regional mining economy and commodification of Appalachian nature. During the late 1850s there was renewed interest in Georgia gold, as California mining professionals attempted to bring new mining technologies to the southern mountains to extract gold inaccessible via previous techniques. The Civil War cut short these efforts but not before the full force of industrial mining began to wash away Appalachian hillsides.

Following the strike at Sutter's Mill in 1848, California mining quickly exceeded the scope of southern gold works. As in Georgia, the earliest prospecting took the traditional forms of panning and sluicing alluvial deposits, but substantial, capital-intensive operations soon dominated California. For all the stereotype of the lone Sierra prospector, the western gold rush was largely a corporate venture. The epitome of California mining came with the introduction of hydraulic mining, which used storage ponds to collect water from fast-flowing Sierra streams, and then flumes and gravity to force that water through massive hose nozzles. When aimed at sloping hillsides, these hydraulic "cannons" condensed the effects of centuries of soil erosion into a few minutes,

THE MONITOR.

FIGURE 3.4. Hydraulic mining involved washing away surface soil and rocks with the use of massive water-powered "guns" like this one shown in operation in California. "The Monitor," Henry Sandham, 1883. *Century Illustrated Monthly Magazine.*

washing away earth and rocks, tearing apart hillsides, and exposing gold deposits unreachable by previous techniques.

Hydraulic mining made California operations more efficient and profitable from the operators' perspectives but quickly led to conflict with downstream landowners in the Sacramento watershed who found themselves buried under the mining waste runoff of the Sierra slopes.[61] According to historian Robert Righter, in the industrial mining techniques of the Sierras "water became the instrument of a business dominated by human greed and unregulated capitalism gone berserk."[62] A decade of hydraulic washing left landscapes almost unrecognizable. The scenes of destruction were so grotesque—yet spectacular—that one, the Malakoff Diggins, would become a California state park. Malakoff Diggins memorialized the folly (and in its case, horrifically ironic beauty) of landscapes washed away by intentional human action and greed.[63]

Miners' prospects for quickly growing rich eroded almost as fast as the mountain soils. Less than a decade after the Sutter's Mill strike, California's goldfields had become crowded, with the largest companies dominating the richest lands. As many fortunes were lost as gained. Some mining engineers began to consider the potential of returning to the old southeastern gold districts with the new technology pioneered in California. They imagined how techniques and machines developed in the face of Sierra nature might work in the forested hills of North Georgia, overlaying real environments with transcontinental understandings of soils, minerals, and hydraulics. To accomplish this, they would first have to bring together speculative capital, engineers, and southern landowners.

A central figure in this swirl of science and speculation (with a healthy leavening of bunk) was William Phipps Blake. Born in New York in 1826, Blake attended Yale University, where he trained in chemistry under the eminent professor Benjamin Silliman. The precocious Blake also developed a keen interest in engineering and geology. He was equally at home patenting a spring-assisted fishing hook as presenting a new theory on continental mountain uplift to a national scientific meeting that included such luminaries as geologist James Dwight Dana.[64] During the 1850s, Blake published a series of geology articles in prominent journals and received a position with the federal government as a part of the geological survey of the new state of California. While working in the state he apparently participated in, or at least closely studied, the hydraulic engineering taking place in the goldfields.[65] After the Civil War, Blake rose to even greater prominence, becoming Arizona's territorial geologist and publishing a number of books on mining, metal ores, and civil engineering.[66]

In the late 1850s, Blake turned to a new line of work: creating a consulting business to exploit renewed interest in the languishing Georgia mines. Operating from his office in New York City, Blake provided engineering advice to Georgia mining operations. His main role in these ventures seems to have been providing prospective investors with assurances that a scientific professional with western expertise backed their risks, and for Blake, the future of southern mountain mining lay in the hydraulic techniques of the Sierras. Typical was his promise "that by introducing water by canals, as in California, and washing down the hills by the hydraulic process, a new era of mining will be inaugurated."[67] This was an opportunity to replicate or even better the success of western mining. One of Blake's tracts promised that North Georgia combined a healthy climate (where even New Englanders would thrive), along with abundant water, timber for building hydraulic flumes, firm soil for canal beds, and—of course—plenty of untapped gold.[68]

This was a science and not a gamble, professionals like Blake asserted. He promised to minimize risk and "change the uncertain, lottery-like character of mining, to the conditions of a regular business, yielding its assured and certain average of monthly or quarterly returns."[69] But he was not above playing on the speculative excitement of gold mining, such as when he flourished an array of gold nuggets from Georgia before a crowd at an 1860 meeting of the American Association for the Advancement of Science.[70] Blake's showmanship and professional endorsements helped raise hundreds of thousands, if not millions, of dollars for speculations like the Yahoola River and Cane Creek Hydraulic Mining Company, Chestatee Hydraulic Company, Auraria Mines, Southern Gold Company, and Nacoochee Hydraulic Mining Company.[71] The Yahoola Company alone managed to sell fifty thousand shares for $500,000 during the late 1850s.[72] Other hydraulic mining outfits also incorporated with investor support during the same years, including the Loud Hydraulic Hose Mining Company, Wood Hydraulic Hose Gold Mining Company, and Cavender's Creek and Field Gold Mining Company.[73]

Blake's work followed a standard template. He would travel south, survey a prospective mine site, and offer some predictions about profitability and how California techniques might be used to great effect in Appalachia. His reports also often included a map or two and sketches of prospective hydraulic canals. Blake's report was then published in pamphlet form in the Northeast, where it was used to drum up capital for the mining operations. Blake had company in this professional-for-hire game. Charles Jackson, assayer for the state of Massachusetts, also sold his expertise as an evaluator of gold to endorse the prospects of various companies with interests in Georgia, as did mining engineer James T. Hodge of Brooklyn, New York.[74] (The outside consultant model would survive the Civil War. For example, New York's Adelberg & Raymond engineering company produced a flurry of survey endorsements in Cherokee, Cobb, and White Counties in 1866 and 1867.)[75] The era's other nascent geological industries also heavily relied on expert surveys and prospectuses to secure investment capital. In southwestern Virginia, such reports attracted northern speculators to lead, copper, and iron mining opportunities. And Benjamin Silliman Jr., the son of Blake's mentor and, like his father, a professor of chemistry at Yale, wrote an 1855 report on the prospects of petroleum in northwestern Pennsylvania that would do much to stimulate the nation's first oil boom.[76]

Blake's stress on the importance of professional, scientific advice reached ready ears, not just in New York boardrooms but also across the South. By the 1850s, southerners were particularly interested in the idea of scientific expertise, most often focusing on reforming agricultural practices. The challenges of an increasingly national and interconnected agricultural economy deeply concerned many southern farmers, especially residents of the seaboard states, where soil exhaustion and emigration were serious problems. Their response was an attempt to intensify and systematize agriculture. Pamphlets akin to Blake's publications (professing expertise on the part of the author and advocating success if only a certain system was followed) flooded the region. From Maryland to Louisiana, planters and farmers formed agricultural clubs, read farming journals, bought new equipment, and experimented with a number of new crop and animal varieties. All of these activities sought a blending of "book learning" and "hands-on" experience. Guides to scientific overseeing were even popular. Blake's mining tracts promised something quite similar: his expertise and methods would redeem old land (in this case, old gold mines), thus generating greater profits than it had ever before realized—if only land-owners and investors would trust him.[77]

Hydraulic mining techniques found their way to Georgia and temporarily revived the state's gold economy, but they did so at a cost. Hoses blasting jets of water against hillsides unveiled more southern mountain gold at the ex-pense of the mountains and valleys themselves. The power of water harnessed by these new techniques was difficult to convey, but Blake described the work succinctly: "The water issuing in a continuous stream, with great force, from a large hose-pipe, like that of a fire engine, is directed against the base of a bank of earth and gravel, and tears it away. The bank is rapidly undermined, the gravel is loosened and violently rolled together, and cleansed from any adhering particles of gold, while the fine sand and clay are carried off in the water. . . . Square acres of earth on the hillsides may thus be swept away into the hol-lows, without the aid of a pick or shovel in excavating." The Appalachian forest posed no obstacle in the face of this hydraulic energy. Indeed, the presence of dense woods made the work more efficiently destructive. "The process is espe-cially effective in a region covered with trees, where the tangled roots would greatly retard the labor of workmen. In such places, the stream of water washes out the earth from below, and stump after stump falls before the current, any gold which may have adhered to the roots being washed away."[78] An outside observer marveled at the power of nature as harnessed by man in this form of mining, for "in this manner the genius and enterprise of our people have met and subdued great natural difficulties; and the streams of the mountains have been made their willing servants."[79]

As Blake saw it, Appalachian nature all but demanded hydraulic min-ing: "The topographical features are more favorable than in the other gold-producing districts of this country, the water being more abundant with a greater power of fall."[80] Even an environmental quality local farmers had long considered detrimental, soils prone to easy erosion, was cast as a stroke of luck for hydraulic miners.[81] Blake well understood the hazards of the methods he advocated—he had, after all, worked in California, where the environmental and social costs of hillside washing were too obvious to ignore. But whether out of a desire for personal profit or a genuine belief that the benefits out-weighed their costs, he brushed aside these worries in his plans for Georgia. Entire mountainsides that had been washed away had to come to an eventual rest somewhere, but fortunately for mining companies "the descent or fall of

FIGURE 3.5. Several Georgia mines, encouraged by the advice of boosters like William Blake, implemented California-style hydraulic mining during the late 1850s. Illustration from William P. Blake and Charles T. Jackson, *Gold Placers of the Vicinity of Dahlonega, Georgia* (Boston, 1859).

the stream[s] [was] sufficiently rapid to carry off the tailings."[82] The gold would remain, and, moving downhill, the debris became someone else's problem.

Hydraulic mining was thus expensive in more ways than one, but the work of experts like Blake, coupled with northern investment capital, brought the California model to Appalachia by the start of the Civil War. Perhaps the first to implement these methods was mine owner H. M. Van Dyke, who began using hoses to wash western North Carolina hillsides sometime in the late 1850s. Van Dyke also worked to transfer the techniques to Georgia.[83] The most significant concentrations of hydraulic mining centered on the gold-rich watersheds of Georgia's Chestatee and Yahoola Rivers and Cane Creek, where the first hose pipes came online in 1858.[84] Foreshadowing twentieth-century Appalachian coal mining, some Georgia mines even began using explosives to destabilize hillsides in preparation for washing.[85]

In the first years speculators made more money from water than gold, constructing flumes and canals to bring water to mining works, where it was sold to mining outfits to operate their cannons. One such water provider, the Chestatee Hydraulic Company, sold hydraulic power to miners in exchange for 10 percent of the resulting gold—and touted their business as a public good. The company claimed that its work made possible, through the power of investment capital, something most small miners could never accomplish on their own: "harvesting the crop of mineral wealth which nature had deposited in the soil."[86] Boston-born Amory Dexter, an engineer and sales agent for the Yahoola River and Cane Creek Hydraulic Mining Company, oversaw the daily work of one such enterprise and kept a diary. Dexter spent much of 1861 walking the company's ditches, negotiating sales of water, monitoring company-owned gold plots for claim jumpers, and maintaining the works.[87] Within just a few years, companies had already filled the Yahoola Valley with tailings from hydraulic mining and "a considerable quantity of auriferous sulphides," although the destruction was not yet on the scale of California's mining wastelands.[88] Modern industrial gold mining had reached Appalachia.

Although the boosterism of men like Blake stimulated renewed investment in the southern mountains, this California dream was short-lived in Georgia because of other national developments. The Civil War brought an abrupt halt to hydraulic mining in the Georgia mountains, severing the connections between northern investors and southern mines and redirecting the South's engineering capacity (along with many miners) into the Confederacy's war effort. The South's desperation for hard currency during the war could not overcome its even greater desperation for manpower. Enlistment fever, the Confederate draft, a dearth of southern industrial capacity to supply mining equipment, and the armies marching through North Georgia later in the war all presented serious obstacles to continued mining efforts. Accurate figures are difficult to come by, but hydraulic mining in Georgia seems not to have reached the production levels of the early 1840s before it was curtailed by the Civil War.[89]

Hydraulic mining and northern speculation would, however, return in several waves after the war's end. During Reconstruction, mining boosters would again tout the mountains as "one of the most inviting fields of a region which, in our judgment, is destined to be largely remunerative to well-directed capital and industry."[90] At the turn of the twentieth century, when new mining technologies briefly promised to make the region's well-worked ground profitable once more, geologists also raised the old lament that "so little development work has been done by the owners and operators of the properties."[91] Georgia's largest mining operation actually formed around 1900, when the Standard Gold Mining Company (an Ohio concern capitalized at $5 million) opened a 120-stamp mill in Dahlonega called the Consolidated Gold Mine. Standard's operation covered five thousand acres, complete with a twelve-mile canal to power its hydraulic cannons and a small ore railway that replaced the traditional mules and carts.[92] But none of these later efforts had any more staying power than the initial gold rush or the late antebellum hydraulic fever. There were simply too many newer, richer global goldfields. By the first years of the twentieth century, Georgia miners had to compete with massive ore deposits in Alaska, Australia, and South Africa, and they failed to effectively do so.[93] By this time too a new mining industry spawned from the same environmental ideas that drove gold mining—coal—had captured the bulk of Appalachian industrial investment.

&

Precious metals mining was one of the central activities of America's nineteenth-century industrialization, and the southern mountains were the epicenter of the early industry, even if California supplanted Georgia after 1848 in practical as well as symbolic importance.[94] The gold fever of the late 1820s and 1830s rapidly populated portions of the mountains with American settlers, in turn promoting a brutal Native American depopulation. Gold strikes accelerated the removal of Cherokee people and connected formerly remote districts to the national economy. They also stimulated an interest in Appalachian mineral resources more broadly, as the search for gold also advanced the mining of iron, lead, copper, mica, coal, and other eventual sources of mountain industry. By the Civil War, there would be pockets of mining throughout the southern mountains, including copper, lead, and iron enterprises in southwestern Virginia, as well as copper mines at Ducktown, Tennessee.[95] In this way, gold mining in western North Carolina and North Georgia started a diverse mountain mining boom that continues to the present.

Mining efforts drew on early-nineteenth-century American conceptions of the wealth and purpose of nature, even as they radically and rapidly changed environments. Some miners saw in North Georgia what William Gilmore Simms expressed in *Guy Rivers*: the mountains were the site of potential riches, yet at the same time they were a wilderness that threatened to undo people's social and intellectual foundations. Nature yielded gold, but in turn it extracted civilization. For other observers, however, the relationship could be reversed. They saw in the hills a sylvan nature despoiled. They romanticized pre-mining nature, often in association with a fanciful Cherokee culture, and lamented that the pursuit of gold—though necessary—was so ugly. The awesome and invasive techniques of hydraulic mining represented the culmination of these concerns about the aesthetic hazards of mining, even as figures like William Blake portrayed them as logical extensions of the expanding fields of hydraulics and engineering.

North Georgia mining was also more persistent than often recognized. Although the California gold rush reduced Appalachia's importance as a center of gold production, regional mining continued, with periodic upswings epitomized by the efforts of capital-intensive hydraulic companies and experts like

Blake. Speculations in the 1850s, during Reconstruction, and again at the turn of the twentieth century periodically revived interest in Appalachian gold and in some instances set capital flowing and miners to work in the ground. Although public awareness of the mountain goldfields peaked during the 1830s, miners worked on for decades longer.

Scholars have identified the late nineteenth century as the era when outside capital and absentee landowners came to dominate Appalachia. As a result, natural resources flowed out of the mountains, and a form of economic imperialism reduced the southern highlands to a peripheral region that fed the growing wealth of the nation's cities. However, we can see elements of this system from the first days of the southern gold rush.[96] Despite the stereotype of the primitive gold panner dominating descriptions of the initial rush, gold mining soon grew into a complex endeavor, reliant on company organization, the wealth of outsiders such as Calhoun and Van Dyke, and production of specie for the national economy. And mining grew even more intensive in the late 1850s with the introduction of hydraulic techniques in North Georgia. This technology linked Appalachia to western goldfields, professional schools like the one at Yale that trained William Blake, pamphlet printers in Boston, and the capital of financial centers such as New York. After the Civil War, gold mining and coal mining looked almost identical in many respects. To take the Standard Gold Mining Company as an example, its Ohio headquarters, high capitalization, railroad building, and promise of strong returns to northern investors read like details of coal boosterism. It is even tempting to equate Blake's lack of concern about the effects of hydraulic mining on downslope property owners with the disregard modern coal operators show for the consequences of pushing the debris created by mountaintop removal into adjacent valleys. Replace hydraulic tailings with "overburden," gold lodes with coal seams, and the stamping mill with coke ovens, and the history of Appalachian gold looks a lot like the more familiar story of the great coal boom.

CHAPTER 4

Salt

Saltville's Civil War

SALT WAS NEARLY AS PRECIOUS as gold in Civil War Richmond. Like many people in the Confederacy, war clerk John B. Jones paid a great deal of attention to the common seasoning and preservative, remarking frequently on it in his diary as the war choked off access to once common supplies. An entry on November 7, 1862, is typical. Jones notes, "Yesterday I received from the agent of the City Councils fourteen pounds of salt, having seven persons in my family, including the servant. One pound to each member, per month, is allowed at 5 cts. [cents] per pound." This increase over peacetime prices was troubling but not as disturbing as the market that had sprung up outside the official channels of supply and price regulation. Jones exclaims, "The extortionists sell it at 70 cts. per pound. . . . Profit $10 per bushel!"[1] As steep as those prices were, the situation would rapidly deteriorate; within two weeks, salt was selling for $1.30 per pound, and Jones was worrying: "We are getting into a pretty extreme condition."[2] Union general George B. McClellan's Army of the Potomac had been temporarily driven from the gates of Richmond, but the city's kitchens and smokehouses still faced a dangerous shortage of salt.

Salt was just as much a daily concern for Confederate soldiers in the field, and sometimes surfeit rather dearth was the trouble. They complained of the monotony of eating salted meat three times a day, as pork and beef so preserved was typically the sole meat for the army. Randolph McKim, a Marylander serving in the Army of Northern Virginia, grumbled that as early as the spring of 1862 "often our only meal in camp was a piece of hardtack and a piece of bacon toasted on a forked stick." He longed for some change in "the unvarying 'menu' of 'slap-jacks and bacon,' or 'bacon and soda biscuit'" that comprised every meal.[3] The tedium of endless bacon was better than no bacon at all, however. As the war ground on, southern rations steadily dwindled until a food shortage became one of the greatest challenges to the Confederate military effort during the final year of the conflict.

Poor though southern rations were, the Confederacy managed to supply its forces in the field for four long years, quite a logistical feat. The flow of food depended on southern farmers and enslaved field hands, tenuous transportation networks, warehouses and rail sidings, a tax-in-kind system, and quartermasters and commissary agents, but the military also relied on an often ignored mineral: salt. In an era before refrigeration or home canning, salt was the dominant preservative, and for the Confederacy the burden of providing that crucial resource fell to a great extent on a single eponymous Appalachian community. Saltville, located in the mountains of southwestern Virginia, became the center of southern salt and would direct the national gaze repeatedly toward central Appalachia as the war wore on. Saltville was a critical source of sustenance for the southern rebellion, and its salt would be another in a long list of mountain natural resources turned national commodities.

Both salt and Appalachia have historically been underappreciated by historians of the war. Salt has long garnered mention in Civil War studies but usually in passing reference. Ella Lonn's *Salt as a Factor in the Confederacy*, first published in 1933, stands as the lone scholarly study of the commodity. (Also worth mentioning, popular historian and science writer Mark Kurlansky includes a chapter on the American Civil War in his *Salt: A World History*.)[4] Likewise, within the grand narratives of the Civil War, Appalachia has often been portrayed as a marginal theater, a bit of no-man's-land between the eastern and western fronts, with the exception of the fighting that took place around Chattanooga and in North Georgia. Relatively recent scholarship has done much to illuminate the Appalachian wartime experience, however, describing a war-torn region of shifting loyalties, guerrilla war, and civilian hardship but also a region more important to the currents of the war than perhaps often recognized.[5] Studying Saltville, the South's central source of salt, which was perhaps the most crucial ingredient in the civilian and military food supply, combines the two subjects and reveals something else as well. Appalachia's natural resources, so important in tying the region to the broader world during the antebellum years, remained nationally important during the war. In some cases the tumult of war amplified rather than cut off the existing commercial connections forged by commodities like fur, botanical specimens, and gold.

Although it lacked the headline-grabbing action of such places as Vicksburg, Gettysburg, or Richmond, Saltville was a vital Civil War site. Politicians and soldiers in both the Union and the Confederacy focused a great deal of planning, troop movement, and fighting around its mines and laborers. Defending the saltworks kept Confederate planners busy and tied down troops and materiel in southwestern Virginia throughout the war. Valuable locomotives, rolling stock, and freight space on the Virginia and Tennessee Railroad carried more salt than soldiers, steel, or guns. Depriving the South of its primary supply of salt likewise obsessed Union strategists, especially as the war wore on, spurring multiple campaigns to wreck the saltworks.

Like antebellum Georgia's gold mines, Saltville is another place to witness early "industrial" forms of Appalachian resource extraction. Salt production was more intensive, more industrial, and more capitalist than is sometimes recognized. It brought together investors' money, state interests, and distant markets while stimulating related industries and drawing enslaved workers into manufacturing. Sociologist Wilma Dunaway argues for its essentially modern, capitalist nature, describing antebellum Appalachian salt production as a good example of how the extraction of natural resources in a peripheral region could become part of, and then drive, complex commodity flows. Iron, wood, coal, slaves, immigrant workers, and consumer goods flowed into the saltworks. Salt flowed out, some transported directly to consumers and some making its way to meatpacking centers, which preserved the pork and beef so vital to national diets.[6] Likewise, historian Randal Hall declares that industrialists in the mid-nineteenth-century Virginia mountains "shared the evolving capitalist framework with other parts of the world."[7] In sum, in the mid-nineteenth century a network of goods and exchanges centered on Saltville, and the war made these connections all the more important.

જી

Humans have known about and used southwestern Virginia salt for centuries. Mississippian people inhabited a village near the present-day town of Saltville long before the arrival of Euro-Americans, and there is archaeological evidence that salt from local saline springs was part of Native southeastern trade networks, connecting area peoples to communities hundreds of miles away.[8] The earliest Virginia colony explorers, fur buyers, and long hunters heard about these salt licks and salty springs by the early eighteenth century, making them waypoints on their trade and hunting routes, and knowledge of these surface salt resources quickly encouraged the search for more substantial underground salt deposits.[9] The licks' location in the New River Valley, whose steep flanking ridges funneled emigrants from the north and east toward the Cumberland Gap and the trans-Allegheny West beyond, soon made the saline deposits a popular stop for frontier-bound settlers and their livestock. And in his *Notes on the State of Virginia* (1785), Thomas Jefferson included the salines in the state's southwestern mountains as one of Virginia's valuable natural resources.[10]

The economic possibilities of the salt district attracted influential figures during the late eighteenth century. At the start of the American Revolution the tract of land containing the most productive salt licks was owned by William Campbell, who gained fame leading rebel forces at the Battle of King's Mountain. His daughter, Sarah, inherited the salines on Campbell's death in 1781, and she married Francis Preston. The Prestons held on to a portion of their property, eventually opening a salt factory in Smyth County, and they also leased the right to make salt on another portion of their holdings to William King, an Irish immigrant. King seems to have been an astute businessman, and his efforts to capitalize the new saltworks attracted investors from outside the region, such as Nashville's Josiah Nichols. Successful commercial-scale salt production in the region dated to the 1790s. As André Michaux, François André Michaux, and John Lyon roamed the southern mountains in search of salable botanical specimens, Preston and King were already manufacturing Virginia salt for distant markets.[11]

Preston's, King's, and smaller wells were located in present-day Smyth and Washington Counties, where past geologic activity had created significant, relatively accessible salt deposits. An immense (seventeen-mile-long) salt fissure stretched northeast to southwest across the region. The most concentrated bed was under present-day Saltville, where a deposit of rock salt underlay five hundred or more acres, and exploratory early boring in the region established the presence of the rock salt, the first such known deposit in the United States (see fig. 4.1). Although saline springs breached the land's surface in places, the rock salt itself was typically found at depths of two hundred feet belowground. This bed of solid salt was enormous, 175 feet or more thick in places. Although the potential profit from rock salt mining stimulated a good deal of speculative interest, the salt bed itself was relatively inaccessible—at least profitably—via the mining techniques of the early nineteenth century.[12]

Saltville's manufacturers could, however, produce salt through other means. The rock salt layer held huge quantities of concentrated salty water—brine— created as groundwater leached the mineral field, saturating the salt bed and filling in fissures and voids, producing what seemed at the time an inexhaustible natural resource. The technology used at Saltville was not overly complicated, but, like gold mining, salt making did require capital investment and a great deal of labor. Workers first had to drill wells, some as deep as four hundred feet, to access the brine. William King's first well was a primitive affair. Hand-dug, some twelve feet wide and two hundred feet deep, and shored with local timbers, it resembled a mine shaft as much as a water well. He soon moved on to more modern boring methods using cast iron well pipes,

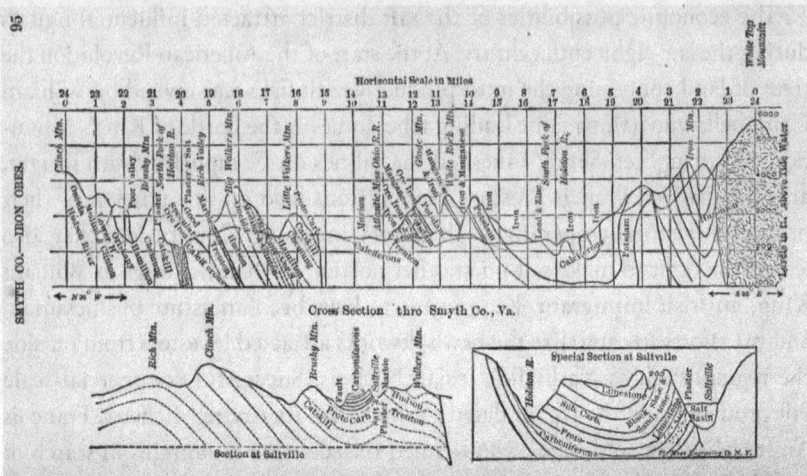

FIGURE 4.1. A cross section of the geology of the Saltville vicinity shows its namesake salt deposits. Charles Rufus Boyd, *Resources of South-West Virginia* (New York: J. Wiley & Sons, 1881).

employing a combination of pumping and siphoning to raise the brine to the surface. Once a well was operational, manufacturers stored the brine in large tanks where it awaited boiling. The evaporative boiling took place either in round metal kettles, or—more commonly at Saltville—in long iron troughs that each held just under one hundred gallons of brine. Constant wood or coal fires in earth trenches under the troughs kept the brine boiling until the water had dissipated and workers could scrape the salt crystals from the evaporators, and a roof enclosed the facility to keep rain from interfering with operations. The resulting salt was then stored in wicker baskets to dry further, before being bagged for sale. Taken together, the well, evaporators, and salt storage spaces represented a primitive factory.[13]

As intense as this production was, expenditures of time, fuel, and labor could have been higher; Saltville brine was exceptionally concentrated. According to boosters, local wells could produce a bushel of salt from 18 gallons of brine, while other antebellum American salines had to boil from 41 to 450 gallons of brine in order to make a bushel. The resulting product was by all accounts of high quality. One observer wrote that "the salt made is free from all impurity, its crystals are large by slow and fine evaporation; and white and brilliant, and when thrown from the [drying] basket, soon becomes as dry as corn-meal."[14] This dry, large-crystal salt stored and shipped well and found a ready market.

Slaves conducted most of this labor-intensive work. Salt makers considered enslaved workers more cost-effective than wage laborers (and though it went unsaid, they likely appreciated a workforce that could be driven harder), and slaves filled most nonmanagerial positions. As was the case in Georgia's gold fields, many of these enslaved workers were tied to the plantation economy. Mine operators leased most of the slaves who worked in salt, either from owners in eastern Virginia, where planters were looking for ways to profit from their "surplus" slaves as the tobacco economy weakened, or from local masters, such as the Floyd family of Burke's Garden, Virginia.[15] This model of enslaved salt labor reached its apogee in the Kanawha Valley mines, north of Saltville, where in 1835 most of the district's three thousand salt workers were slaves. (The importance of black labor persisted after the war, as Booker T. Washington—himself briefly a young salt worker in Malden, West Virginia—famously documented in his autobiography.)[16] Regional slave leasing for industrial work was common. The lead, copper, and iron works that developed in southwestern Virginia as well as the Virginia and Tennessee Railroad also relied on the efforts of enslaved people.[17] Akin to Appalachian gold mining, salt making became both an industrial enterprise and an extension of southern slave agriculture.

Like so many southern plantation owners, salt makers and boosters also lobbied for better internal improvements to rush their products to market. The topography and geology of southwestern Virginia had created a valuable resource, but they also presented challenges to transporting and marketing it. According to one historian, "the first internal improvement in southwestern Virginia was a toll road from Saltville to Abingdon, built in 1803," mainly for the use of salt wagons from the Preston and King works.[18] Observers called for more such improvements. An 1834 article argued that better roads, or a branch of the James River and Kanawha Canal, would lead to an expansion of the salines and make the surrounding countryside prosperous.[19] And in 1843 the northern *Merchants' Magazine and Commercial Review* declared the Saltville region, out of all the emerging salt works from Michigan to Mississippi clamoring to attract investors, to be the site with the greatest potential.[20] The ultimate "improvement" came in 1856, when the Virginia and Tennessee Railroad—connecting Lynchburg to Bristol—was completed (by 1858 it extended to Knoxville). The new line ran the length of the New River Valley and connected to tracks that led to Georgia, Mississippi, Alabama, and the Mississippi River at Memphis. By 1861, the Virginia and Tennessee owned more locomotives and more boxcars and flatcars than any other rail line in Virginia.[21]

Thanks to new infrastructure like the railroad, despite their relative isolation even the relatively small works at Saltville could produce exceptional

Making Salt, at Saltville, Virginia.

FIGURE 4.2. Saltville manufacturing relied on enslaved labor for making salt. "Making Salt, at Saltville, Virginia," illustration from a sketch by J. Wells Champney, in Edward King, *The Great South* (Hartford, Conn.: American Publishing, 1875).

profits. In 1808, after just over a decade in business, William King had an estimated fortune in excess of $1 million. This wealth came with relatively low overhead costs. At the peak of southwestern Virginia salt production in the 1830s, King's works kept approximately one hundred laborers and an equivalent number of horses at work.[22]

But this wealth came with costs besides those for labor, what an economist might label "externalities." Commercial-scale salt making carried environmental consequences. The most dramatic was the deforestation of thousands of acres surrounding the salt works, as the work of evaporation demanded abundant supplies of fuel, consuming one natural resource to produce another. Before mineral coal became the dominant energy source, cordwood and charcoal from local forests fed the evaporator flames. William King's operations alone consumed an estimated ten thousand cords of wood each year by the 1850s, and, when the Civil War began, Saltville's supplies of cordwood came from forests as distant as thirty miles. This timbering represented an early market use of Appalachian forests, predating the more famous post–Civil War logging boom, and most woodlots near salt works had been cut two or even three times by mid-century.[23] Salt making stimulated not only logging in southwestern Virginia but other industrial activity too. Local iron mines grew in part to supply the evaporating pans and machinery needed at Saltville, and regional coal mining, though remaining small in scale, developed to provide an alternative fuel to wood.[24] When a traveler submitted observations on the industry of the local region to *Farmers' Register* journal in 1834, he noted that in the vicinity of Saltville there were "twenty iron making establishments" in addition to abundant quantities of lead, "stone coal[,] and innumerable seats for water-power."[25] Each of these ancillary industries also denuded forests.

The necessity of salt in an age before refrigeration and the location of Appalachian salt makers between the livestock-rich transmontane West and the rapidly growing markets of the Eastern Seaboard briefly positioned Appalachia's salines as a vital force in national development. Indeed, Frederick Jackson Turner, in his influential essay "The Significance of the Frontier to American History," credited the salines with shaping American backcountry settlement and commerce. Only with the discovery of the minerals at Saltville and similar sites, he wrote, had the West begun "to be freed from dependence on the coast. It was in part the effect of finding these salt springs that enabled settlement to cross the mountains."[26] By the mid-nineteenth century, however, Saltville's day seemed to be passing due to a confluence of forces. The discovery of new salines along the western frontier, the increase of salt production in western New York spurred on by the opening of the Erie Canal, repeal of a protective tariff on foreign salt, and the movement of the frontier farther and farther from Appalachia all worked to lessen the importance of southern mountain salt. In 1810, central Appalachia's producers had led the nation in salt production, but by the 1850s they had fallen well behind New York facilities.[27] The Civil War would drastically elevate the importance of Saltville's products, however, temporarily restoring the region's economic significance.

When war finally came in the spring of 1861 after more than a decade of continuous sectional conflict, salt was an immediate and pressing worry for Confederate officials. The necessity of an abundant supply of domestic salt stemmed in part from southern diplomatic and economic plans. Many Confederate politicians believed that the South's leading role in global cotton production would certainly result in disproportionate political power, but prognosticators also argued that should King Cotton fail to force Britain or France to intervene on behalf of the Confederacy, then the South's vast agricultural landscapes could be turned to greater food production. This would eliminate the Deep South cotton districts' antebellum dependency on midwestern farm products (especially pork from meatpacking centers like Cincinnati) and provide a supply resiliency that ought to make the new nation all but unconquerable. A vital link in this projected increase in southern bacon was salt. Without the mineral needed to preserve pork and other meats, larger herds of hogs could provide only local, short-term benefits. If southern soldiers were to march into battle on full stomachs, the Confederacy needed to have a dependable supply of salt.[28]

It seemed likely that this salt would have to come from Appalachia. Only two significant centers of salt production existed in the seceding states—Virginia's works in the Kanawha Valley and those at Saltville—and the Kanawha salt wells were quickly lost to the South when Union forces under Major General McClellan pushed into western Virginia from Ohio in the summer of 1861. In the early months of the war, state and Confederate agents would explore the potential of other salt licks and saline springs across the South, in particular noting a promising deposit of rock salt at Avery Island, Louisiana (although it too would soon fall behind enemy lines). The Confederacy also stressed frugality and enterprise in wresting consumable salt from nature. Officials encouraged coastal residents to evaporate seawater and established coastal salt works, though they faced near-constant threat from the Union navy. Enterprising civilians even engaged in digging up the dirt floors of smokehouses in order to leach out the brownish-gray, odiferous salt that had fallen from years of curing meat. Reflecting fears about short salt supplies, descriptions of alternative (but relatively ineffective) means of curing meat would fill wartime publications.[29]

Despite these efforts, Confederate officials worried that salt shortages would hamper the war effort. The rebel government debated making an exception to its policy prohibiting all trade with the Union if such commerce would bring salt into the South, but it ultimately rejected this.[30] At the end of the conflict's first year, Frank Ruffin, a Confederate commissary of subsistence, wrote to his superior officer of the challenge of salt procurement. The Confederacy needed salt pork, but it had relatively little existing infrastructure for slaughtering, salting, and packing meat. Ruffin estimated that in the last year before the war roughly three million hogs had been salted and packed in the United States, but less than twenty thousand of those had been processed in the South. It was crucial, therefore, to secure salt on a massive scale, and Ruffin asserted that Saltville was the key "for army supplies and for packing purposes."[31] If the salines of southwestern Virginia were lost or unproductive, the rebellion might be doomed.

Ruffin's concerns were well founded. Salt shortages became one of the vexing political and material issues challenging the cohesion of the southern states and testing the central authority of the Confederate government as the war progressed. As early as the fall of 1862, Mississippi governor John Pettus wrote to Confederate president Jefferson Davis that Saltville's production (or at least ineffective distribution of the salt produced) was unable to supply Deep South demand, and "many of our people . . . now have no meat and have had none for many weeks, because they have no salt to season fresh meat with."[32] Alabama governor Jonathan Shorter lodged a similar complaint. Despite paying or promising nearly a quarter of a million dollars to Saltville producers, he fretted that "what is shipped is taken by the State of Virginia." As a consequence, "The salt famine in our land is most lamentable."[33] Nor were Virginians spared. Famed diarist Mary Chestnut, writing in the Confederate capital, expressed a common civilian complaint when she griped about the price of salt and salted meat in Richmond's markets in late 1863.[34]

The state most plagued by persistent salt shortages over the course of the war and most disturbed by Virginia's control of the crucial resource, however, was North Carolina. Saltville's supply was so close to North Carolina's border, yet entirely beyond its political control. The state attempted to address its lack of salt in a number of ways. Most extensive was an effort to build a state facility for evaporating seawater on the Atlantic coast. The government first established saltworks at Morehead City, which were soon captured by the enemy, and then it erected new works near Wilmington. There too successful salt making proved elusive, as an outbreak of yellow fever shut down production almost as soon as it began. North Carolina also exempted salt workers from the draft as vital industrial employees and put at least a few Quakers who were conscientious objectors to work in the evaporative facilities. The coastal works were capable of producing more than sixty thousand bushels of salt per year, but this was not nearly enough to meet state demand, and so the North Carolina government maintained an interest in a share of Saltville's production.[35]

North Carolina, along with other southern states like Mississippi and Georgia, set up its own saline works at Saltville through an arrangement negotiated with Virginia. North Carolina's facility manufactured the bulk of the state's salt supply. Laboring around the clock, workers at its two hundred kettles could produce as much as three hundred thousand bushels of salt each year.[36] But even that quantity was not sufficient for the state's needs, especially when supply was interrupted, as in the snowy winter of 1862–63, when treacherous roads made transporting salt into the mountains of western North Carolina particularly difficult.[37] Deprived citizens were especially disgruntled that they lacked the necessity when an abundant supply existed so close by. Ashe County's Jesse Reeves wrote to the state's governor that winter, "We have a large surplus of hogs in ashe if we could get salt to save them." If the North

Carolina government would not satisfy demands, he suggested that the result might be direct action, noting that "thare is thousands of bushels at Saltville Va & we cant get it & we the Citisons of ashe has com to the conclusion to go & take it by force if the oners of it wont let us have it for a fair price . . . we are compeled to have salt while it can be had or we will fite for it I can get from 3 to 4 hundred men in one days notice to starte [toward Saltville]."[38] Reeves threatened a very different sort of civil war, and at least some other North Carolinians sympathized with his frustration.

Feeling pressure from the state's citizens and habitually irritated by the expanding power of the Confederate national government, Governor Zebulon Vance, himself from western North Carolina, lobbied for a more-secure salt supply. In the fall of 1864, upon learning that Virginia officials had conscripted slaves from North Carolina's saltworks to labor on Saltville's defensive lines, he wrote to Confederate secretary of war James Seddon. Vance pleaded with Seddon concerning such conscription, "Can you not forbid it? . . . [F]or God's sake don't deprive this whole community of the means of living for the sake of forty men." He then turned his anger to Virginia's governor, William Smith, initiating a venomous exchange over Virginia's management of the saltworks and the logistics of the railroads and rolling stock that moved most of the supply to the rest of the Confederacy. North Carolina was barely getting enough of the crucial resource, Vance bitterly complained, in large part because of Virginia's actions that benefited the commonwealth at the expense of its neighbors. Vance seemed to imply that such a vital natural resource should be nationalized, at least during the emergency that was the war. It was an understandable stance in light of North Carolina's desperate need for salt, yet it was notably at odds with Vance's broader concerns about state sovereignty.[39]

For its part, Virginia's wartime government sometimes seemed as troubled as blessed by the South's best salt supply. Governor Smith, in an address to the Virginia General Assembly, sighed, "It is very much to be regretted that this gift of God to man cannot be enjoyed in peace. The States of North Carolina, Georgia, Alabama, and Tennessee, through their agents, complain of their treatment." He did not, however, want Virginia to monopolize the abundance of nature and in doing so be labeled miserly. "It will not do for Virginia," Smith declared, "controlling as she does this great necessity of man, to allow such a state of things as may even plausibly subject her superintendent to the imputation of using the great power which must be confided to him for his private ends."[40] Still, Smith recognized the economic and political advantages of controlling Appalachian salt, and he was determined that Virginia harvest these fruits of natural advantage.

In his direct correspondence with Vance, Smith was just as irritated but less noble. He responded to Vance's complaints about the commonwealth's railroad management by noting that despite its sovereignty over the valuable mines, Virginia received only a fraction of the total salt produced, and its sorely taxed rail network bore a disproportionate share of the freighting. He acerbically reminded Vance that the commonwealth's landscapes and population had seen the brunt of the fighting, resulting in "a country ravaged by the enemy and a people stripped in thousands of cases of the last mouthful of food to all the horrors of want and famine." These challenges were made worse by "the Confederate Government having so pressed upon our transportation as to prevent the distribution of salt among our people," with some of that burden coming from efforts to supply salt to other states (such as North Carolina). In conclusion, Smith suggested that if Vance felt unsatisfied with Virginia's management, North Carolina might look elsewhere for its salt. It was a dire threat indeed.[41]

Virginia moved to secure its interests in Saltville as the war wore on, asserting control over the private manufacturers who had continued to operate in conjunction with the various state works. To make production more efficient, a state superintendent was put in place to oversee salt making and distribution on-site. In addition, the state impressed the Stuart, Buchanan, and Company saltworks and eventually in 1864 assumed direct control of all saline works in southwestern Virginia, declaring them vital to the continued prosecution of the rebellion.[42]

If consumers and their state political representatives were worried about shortcomings in the production and distribution of salt, the Confederate government was even more concerned that Saltville might be lost to Union military action. Early in the war and after Kanawha's occupation, Confederate commissary officials identified Saltville as a crucial producer of military supplies, and troop positions and movements in central Appalachia were often planned with an eye toward protecting the salines.[43] As the war progressed, Jefferson Davis himself maintained an interest in protecting Saltville, emphasizing its importance in an address before the Confederate Congress and personally furnishing military instructions when the saltworks were threatened by a Union raid in force in late 1862.[44] Davis's worries were exacerbated by department commanders' regular insistence that Saltville was "the vital point of this section of the country" and a natural target for Union commanders who intended to undercut the Confederate war effort by drying up its supply of brine.[45]

Civilian and political wrangling over salt guided a great deal of troop positioning in Appalachia throughout the war. Soldiers marched, counter-marched, and threw up defensive works over the course of several years, with an eye toward defending the vital resource. From the conflict's first months, when the Kanawha works were taken and it became clear that Saltville would be the Confederacy's main producer of the mineral, southern troops remained at the salines or close by. When one unit was drawn away, another repositioned to take its place. Often Confederate units that had seen combat in other theaters of the war would be stationed at Saltville on guard duty while they convalesced. Such was the case for the Fifty-First Regiment of Virginia infantry, which fought in eastern Virginia and the Shenandoah Valley but spent considerable stretches at Saltville protecting the works from bushwhackers and potential Union raids.[46]

Periodic reports of Union troop movements near the mountain passes along the Virginia and Kentucky border prompted great concern among Confederate leadership. One such moment took place in December 1862, when Brigadier General Samuel Carter led a federal raid into East Tennessee with the intention of cutting the Virginia and Tennessee Railroad and providing encouragement to mountain Unionists who had been appealing to President Lincoln for aid. As important as the rail line was for moving southern soldiers and supplies, Confederate brigadier general Humphrey Marshall worried that the Union push toward northeastern Tennessee was but a feint. In a summary report to Richmond, he wrote that there were as many as four thousand federal cavalrymen in the vicinity of the Cumberland Gap, and "the salt works in Smyth County (18 miles above Abingdon and near the railroad) would be an interest at which such a force would be most apt to direct its efforts. It might make a feint upon Bristol (an unimportant depot) while the main body might move rapidly upon Saltville, and in an hour might there do damage that would be nearly irreparable."[47] The railroad might be important, Marshall argued, but the salt supply was crucial.

Carter's raid was deflected, and careful movement and positioning of Confederate defensive forces continued the following year. Infantry and artillery units remained garrisoned at the salines throughout 1863, and the commander of southern forces in southwestern Virginia, Brigadier General William Preston, argued for the independence of his command from the Army of Tennessee, first and foremost to ensure that he could devote full attention to defending the salt works. He stressed that "the main vulnerable point here is Saltville, which produces 10,000 bushels of salt per diem, and which is of vital consequence to the Confederacy."[48]

If nature had created a crucial resource at Saltville for the new Confederate nation, it also presented great challenges to defending it. Southern commanders noted that although the mountains north and west of the salines formed a defensive obstacle of sorts, their many passes also gave Union raiders multiple avenues of attack, and every report of troop sightings near the Kentucky line brought a fresh round of worries for Saltville's defenders.[49] On multiple occasions in Richmond in 1863, clerk John Jones heard of Union action near the Cumberland Gap and worried that Saltville might fall and with its surrender lead to a collapse of Confederate food supplies.[50] Often these reports proved nothing more than combined mistakes and paranoia, but occasionally they contained some substance, as in the winter of 1863, when a Union officer in pursuit of General James Longstreet after a Union victory at Knoxville announced his intention to cross the Holston River, anticipating "a favorable occasion for making a dash at Saltville with the whole mounted force."[51]

In addition to ensuring a constant garrison, the Confederacy worked to further strengthen the salines' defensive works. In the summer of 1863, Major General Samuel Jones impressed slaves from surrounding counties to build earthen fortifications, and, conflicting with the South's extreme demand for salt, he also temporarily conscripted salt manufacturers' slaves to assist in constructing the defensive lines.[52] The crucial nature of defending Saltville was emphasized by the army's decision in March 1864 to create a new military organization, the Department of Southwestern Virginia, with the primary responsibility of defending the salt supply.[53]

The general concerns about Union intentions toward Saltville were well founded. In late winter of 1864, General Ulysses S. Grant—by that time in overall command of Union efforts in the field—instructed Major General John M. Schofield to make every effort to capture or destroy the saltworks. If successful it would be a major blow to the South, especially as the campaign might coincide with Grant's summer plans in Virginia and General William Sherman's movement south into Georgia from Chattanooga. Destruction of the "salt-works there," Grant wrote, "would compensate for great risks."[54] Likewise, General Stephen Burbridge attempted to motivate his troops on campaign in southwestern Virginia by asserting "that depriving the Confederates of the saltworks would do more to bring down the Confederacy than the capture of Richmond."[55] Saltville had become nearly as important for the Union war effort as for Confederate plans.

By the time of Burbridge's speech the Union military was in position to launch a more concerted campaign to capture Saltville. As noted, Grant encouraged such efforts in February, with the result being a failed raid in the spring.[56] A more sustained campaign took place in the fall, led by Burbridge. Its culmination was a pitched engagement—the Battle of Saltville—that occurred on October 2, 1864. A force of forty-two hundred Union soldiers, led by the First Division of the District of Kentucky (composed in part of African

American troops), forced Confederate defenders under the overall command of John C. Breckinridge to fall back into interior defensive lines. Despite the fact that the Yankees "repeatedly charged the earth-works with their guidons flying, [and] suffered considerable loss," the defenses held.[57]

The Union suffered approximately 350 casualties in the battle, but it was violence after the action that captured national attention. Here as elsewhere in the conflict, African Americans were an integral part of the war effort. The Confederacy had conscripted enslaved salt laborers to construct defensive works, and free black Union soldiers assaulted those same works in an effort to help end the institution of slavery, bringing even a corner of Appalachia perceived as relatively remote into the national struggle over enslavement and freedom.[58] And those efforts drew the ire of many white southerners who saw in African American soldiers the personification of their fears of a race war. Confederate soldiers under the command of Brigadier General Felix Robertson reportedly killed a number of wounded black Union men in the immediate aftermath of the battle. And southern partisans led by guerrilla Champ Ferguson entered a hospital at nearby Emory & Henry College on October 7 and 8, executing a number of injured black soldiers and a white officer, and threatened more murders in retaliation for the North's use of African American combat troops.[59]

Coming only a few months after news of similar executions of black prisoners at Fort Pillow in Tennessee, the "Saltville Massacre" further raised northern ire, with papers such as William Lloyd Garrison's *Liberator* declaring it additional evidence of "rebel barbarities."[60] For his part, Burbridge declared the executions conducted by Ferguson's men in particular "one of the most diabolical acts committed during the war." He pledged that if captured, Ferguson and his accomplices "will not be treated as prisoners."[61] Burbridge's failed drive on Saltville and its bloody aftermath kept the strategic importance of Appalachian salt on southern and northern minds. Southerners breathed a temporary sigh of relief that the supply remained safe, while Union strategists crafted a new campaign to deprive the Confederacy of salt, one perhaps further fueled by the racial violence that accompanied the previous effort.

Union high command settled on a quick-moving raid in force to capture the crucial production center. The idea for a new Saltville raid came from General George Stoneman almost as soon as Burbridge's campaign failed. Stoneman, a cavalry officer, had a less-than-stellar war record by the fall of 1864. At Chancellorsville the year before, he had led a raid that reached the outskirts of the Confederate capital, but his separation from General Joseph Hooker's main army, followed by Hooker's resounding defeat at the hands of Robert E. Lee, called Stoneman's decisions into question. As part of Sherman's

FIGURE 4.3. General George Stoneman, between 1861 and 1863.
Courtesy of the Library of Congress.

Atlanta campaign in the summer of 1864, he again separated his command
from the main body of Union troops and directed an ill-conceived raid on the
Andersonville prison. The effort fell apart at Sunshine Church, Georgia, where
Confederates took Stoneman as a prisoner of war (he was the highest-ranking
Union officer captured in the Civil War). After being exchanged in September and seemingly unfazed by his previous failures at long-distance cavalry
raids (or perhaps seeking to redeem his reputation with a successful example),

Stoneman proposed a new drive on the works at Saltville and the nearby lead mines that were also crucial to the Confederate war effort.[62]

Collectively, these Union plans for southwestern Virginia (and most critically for the salt fields) amounted to an assault on nature as much as on the Confederacy. Lisa Brady has shown how federal actions in 1864 and 1865—especially Philip Sheridan's Shenandoah Valley campaign and Sherman's march through Georgia and the Carolinas—essentially amounted to a "war upon the land." Union strategy increasingly targeted southern farmland, the nexus of nature and culture that might be termed an agro-ecosystem, and especially the farm technologies like fences, barns, and agricultural machinery needed to produce food. This strategy was an effort to lessen the South's productivity and weaken civilian support for continuation of the war.[63] Raids on salt, lead, and iron works in the mountains grew from much the same purpose, targeting natural resources and the infrastructure that exploited them, essential for both war making and civilian life.

Stoneman would achieve what previous Union forces could not. His raid finally broke through Saltville's defenses on December 20, and Union soldiers did their best to ensure that the salines could no longer aid the Confederacy's war efforts. Troops ransacked the nearby lead works at Wytheville, tore up sections of the Virginia and Tennessee Railroad that moved salt and soldiers through the region, and destroyed salt-making sheds, kettles, and furnaces in Saltville itself.[64] The raid also captured between fifty thousand and one hundred thousand bushels of salt, so desperately needed across the South in the winter of 1864–65. In his after-action report, Stoneman crowed of "the total destruction, as far as in the power of man to accomplish, of the lead-works, seventeen miles from Wytheville, and the salt-works at Saltville. The furnaces, kettles, and machinery were broken into pieces, the wells and shafts filled up with shells, railroad iron, &c., and the buildings burned down."[65] Northern papers likewise celebrated the victory and subsequent destruction, confident that the loss of the salines spelled the end of the Confederacy. The *New York Observer* labeled the raid a crucial blow to the South, and the *Zion's Herald and Wesleyan Journal* called it a setback that could not be "repaired during the war."[66]

The damage did seem to spell the end of Confederate salt manufacturing, at least in southwestern Virginia. As one later economic historian would note, Stoneman was correct in claiming the "almost complete destruction of [Saltville's] plant and equipment."[67] The Confederacy, however, was determined to renew manufacturing. From within the War Department, John Jones recorded Saltville's capture but calmly noted that the Union failed to occupy the site and that thus the works could be repaired.[68] Salt makers' initiative and the

Confederacy's desperate need for the mineral led to a resumption of wartime production, even if it could not regain the same scale as earlier. Supplies of salt continued to trickle from the salines until the South's surrender.[69]

Just as the firing on Fort Sumter renewed Saltville's prominence, Lee's surrender at Appomattox immediately lessened its importance. With the end of the war the mines lost their crucial position as the central supplier of salt to a nation, and they once again faced the competitive forces that had been undermining Appalachian salt's economic position during the 1850s. Despite losing their competitive advantages, southwestern Virginia's salines would continue production for a time. Edward King, a journalist for *Scribner's Magazine*, visited Saltville in the 1870s as part of his extensive tour of the South and described "a neat manufacturing village" that had erased all evidence of Stoneman's raid. Coal rather than cordwood now fired the evaporation kettles, the town also specialized in mining gypsum for agricultural markets, and the hard labor of both industries continued to largely be the work of African Americans.[70] Another observer would claim that the local mineral deposits remained an "inexhaustible resource," but such boosterism was rampant in the postwar South, especially in Appalachia.[71] By the late nineteenth century, Saltville mines would in fact exceed their antebellum production levels, but the mining was increasingly capital-intensive and decreasingly important as a percentage of national production.[72]

℘

Appalachian salt during the Civil War provides another example of the ways in which regional natural resources forged connections between the mountains and the lowlands. From the time of the American Revolution until Fort Sumter, Virginia salt was a valuable commodity, but its production was relatively small-scale, and the salines were but one of a group of industries in the southwestern corner of the state that included lead, copper, and iron making. On a national scale, antebellum salt illustrated much the same connections and dependencies as did the fur trade, the market in plant specimens, and the Appalachian gold rushes. Mountain geography posed obstacles to settlement and development, but demand for mountain resources simultaneously worked to overcome those environmental challenges. And labor as well as consumption connected salt making to people outside the mountains. Relying in many cases on leased slaves to make salt, the salines were integrated into the southern plantation economy in much the same way as was Georgia's gold mining.

Saltville's nineteenth-century history is also a reminder that there was a great deal of human geography at work in the Appalachian Mountains. The Civil War temporarily redrew political space in a way that changed the value

BROWN HILL FURNACE, WYTHE CO., VA.

(P. 64.)

FIGURE 4.4. Brown Hill Furnace iron works, Wythe County,
Virginia, in Charles Rufus Boyd, *Resources of South-West Virginia*
(New York: J. Wiley & Sons, 1881).

of the products of nature, and no commodity made this more apparent than salt. Salt was essential for the Confederacy. It kept its soldiers and civilians fed, and without it the ability to fight would have rapidly dissolved. Politics and geology intersected in the location of Virginia's salines to temporarily elevate Saltville from a relatively minor site of national production to perhaps the Confederacy's most important center of resource extraction during the war. There nature and culture met at the brine wells.

Civil War historians and buffs are also rediscovering the importance of salt in guiding wartime decision-making. In part this is an outgrowth of the nascent subfield of Civil War environmental history, as scholars explore the natural world's influence on the conflict as well as the ways war shaped ideas about material environments.[73] And, as historian Stephen Berry notes, this new environmental focus requires a renewed examination of the flows of natural resources during the war. As a consequence, the study of both logistics and environments are increasingly popular, with Ella Lonn's *Salt as a Factor in the Confederacy* (1933) finding renewed scholarly appreciation.[74] On the popular front, some local history buffs and even reenactors have found salt an engaging window into wartime material culture and military action. Living history events and new historical signage at Saltville as well as locales as far afield as Swansboro, North Carolina; Charleston, West Virginia; and Panama City, Florida, have all recently dealt with battles at or raids on Confederate saltworks.[75]

Environmental historians of the war have to this point focused on the importance of disease in the conflict, efforts to modify the environment for tactical or strategic purposes, and the destruction that combat wrought upon nature as well as people. A close look at salt reveals the important ways in which considerations of nature—beyond the classic studies of battlefield topography—guided wartime decision-making. Confederate and Union notions of the importance of Saltville's salines, coupled with their location, influenced the movement and actions of tens of thousands of soldiers across multiple campaigns. The dearth of southern salt and the singular location of most of its manufacture also led to civilian worries and altered diets, influenced the management of the Confederacy's precious freight trains, led many a farm wagon to trek the mountain roads to Saltville, and kept complaining letters flying between state and Confederate officials. In almost every corner of the nation between 1861 and 1865, people were thinking about the importance of a once common mineral and the relatively remote Appalachian saltworks that produced it.

CHAPTER 5

Transportation

Roanoke, Railroads,
and Appalachia on the Move

ROANOKE, VIRGINIA, is a city still defined by railroads. If you fly into the commonwealth's western commercial hub, nestled in the Roanoke Valley between the Blue Ridge Mountains to the east and the long ridges of the Alleghenies to the west, as your plane circles to descend into the regional airport it passes over the sprawling yards of the Norfolk Southern Railway. Miles of tracks, lined with sidings, empty coal cars, warehouses, and machine shops, transect Roanoke on their paths connecting the Eastern Seaboard to the Appalachian Mountains. Railroad materials and auxiliary facilities clutter the best valley real estate. Viewed from an automobile the geography of railroads is just as prominent. The Interstate 581 corridor that bisects downtown crosses over the rail yards, allowing views of trains and tracks from a closer vantage and reminding commuters of the city's economic past and present.

As with railroads, the city's relationship with nature is noticeable regardless of mode of travel. Passenger planes landing at Roanoke often make a dramatic circle of the city because the Roanoke Valley is ringed with mountains that hinder a more direct approach. Cars entering town from the south on U.S. 220 thread their way along Maggodee Creek and Back Creek in the shadows of hollows as they pierce the Blue Ridge (crossing under the stone-clad bridges of the Blue Ridge Parkway) to enter the valley. From the west, interstate traffic descends the mountains between the city and Blacksburg and then passes through a pastoral landscape of agricultural land and subdivisions framed by mountains in all directions. In the city itself, almost every view is backgrounded by the bulk of Mill Mountain, illuminated at night by its distinctive giant neon star near the crest of the ridge.

Another sort of intersection, a cultural and intellectual one, marked Roanoke's early days. The city's founding in the 1880s and its explosive growth in the following twenty years mirrored dramatic changes taking place across Appalachia at the time. Roanoke represented two diverging ideas of the region, ideas that had roots in the same societal and economic forces. On the one hand, the city epitomized a modernizing, increasingly urban nation. Across the United States following the Civil War, towns blossomed into cities, and cities grew into metropolises, fueled by high birth rates, more industrial jobs, and heavy immigration. In its expansion from a minor crossroads to a regional economic center dependent on rail lines in the 1880s and 1890s, Roanoke was prototypically American.[1] On the other hand, Roanoke's boom also seemed atypical. The city rose to prominence at the moment Americans were coming to think of Appalachia as an exceptionally backward, rural place, a land that time forgot.[2] In the last decades of the nineteenth century, local color writers wrote reams of pages about the remote "hollers" of the southern mountains; collectors sought out authentic handicrafts and old ballads they believed had roots in the distant past; charitable social reformers sought to uplift and reform mountaineers' morals, hygiene, religious habits, and linguistic quirks; and boosters defined Appalachia as a landscape of chronic underdevelopment, even more in need of outside investment than the rest of the destitute post–Civil War South.[3] In this increasingly "other" region of America, Roanoke's race toward modernity seemed anomalous and therefore worth noting. An examination of the city's economic base reveals, however, that Roanoke's growth relied on much the same thing as many other American cities booming during the same years: the extraction of natural resources and the needs of transportation technology. How Roanoke's development played out, however, stems from the particular natural resources of the mountains surrounding it.

This intersection of mountain geography with technology in the form of railroads has defined Roanoke from its early days and its birth in the Appalachian coal boom. Roanoke is one example of the Appalachian urban explosion that occurred at the nexus of mountain resources and the voracious demand of industrial centers nationwide. Roanoke grew to be a waypoint, a town and then a city that funneled the products of the coalfields and Appalachian laborers through its rail hub and warehouses and then distributed them widely to the broader nation, amassing wealth and regional influence in the process. Roanoke was a city of workers and merchants who dealt in two commodities: coal, so central to an increasingly industrial nation, and the motive power in the form of railroads necessary to transport it. As with Chicago's connection to the Great Plains, midwestern prairies, and the piney woods of the upper Great Lakes—made famous by historian William Cronon's study—Roanoke was a place made by its connections to other places, a city that linked commodities and consumers.[4] Whereas Chicago had its connection to the Great West, Roanoke was a gateway to the Pocahontas coalfield. Between the city's mountains and the rail yards it is hard to ever forget that the products of nature played an important role in the rise of Roanoke.

Roanoke is the best place to witness the rise of urban pockets of Appalachia, as it is in some ways a typical case and in others the most extreme. Within the mountains no site went from rural countryside to important city more rapidly than Roanoke: it was the classic boomtown, even as the forces populating it represented broader regional developments. And, as the operational hub of the Norfolk Southern Railroad, no city relied more on its railways, the transportation form that made possible the exploitation of Appalachian natural resources on a new scale. This transportation itself became one of the region's most important commodities, exploiting geography and dividing distance and weight into units of measurement for which companies could charge. This commodification of movement in turn relied primarily on another booming post–Civil War commoditized natural resource: coal. Along with the importance of commercial timber sales across much of the southern mountains, coal and railroad haulage formed a triumvirate of commodities that came to dominate regional economies. With its location in a valley between what was for a time the nation's most important source of coal and its mass of East Coast consumers, Roanoke existed at the center of a rapidly changing late-nineteenth-century Appalachia. And urban pockets like Roanoke, rather than being exceptions to a rural, hardscrabble region, were expressions and concentrations of its increasing reliance on the transformation of mountain nature into commodities.

ꙮ

"Big Lick," as the community that would become Roanoke was called from its murky origins in the early 1800s until its boom at the start of the 1880s, was a literal stop in the road. Much like Saltville to the west, salines put Big Lick on the map. Its site on the swampy bottomland where several creeks came together included a salt deposit that attracted deer and buffalo. This abundant game in turn drew Native Americans and then colonial hunters engaged in the transatlantic fur trade. Following the game and hunters' trails, settlers heading west toward the Cumberland Gap or south on the Great Wagon Road used the lick to satisfy their livestock on the journey through the mountains. Some hunters and salt collectors noted that local land looked promising for agriculture, and over time a few established farms in the area. One early visitor wrote in 1795 of passing "many excellent Plantations some of wch had extensive Bottoms of Rich Land upon the Roanoak Creek."[5] The Virginia and Tennessee Railroad reached this quiet farming community from the east in 1852 (it would be completed to Bristol, on the Tennessee line, by 1856), but without regional mining or large-scale timber operations the railroad did little to promote growth in the small crossroads. A stop on the wagon road became a stop on the rail line.[6] As late as the 1870s the farm town of Salem, located a few miles west of Big Lick, remained the more important hamlet, if only for providing a modest business in storing and loading farm goods from surrounding counties for shipment on the railroad. Farmers complained that their goods had to be brought through the mountains on a carriage way that was a "terrible rough ride of thirty miles."[7]

But travelers and investors turned a keen eye on the South after the Civil War, seeking opportunity in the turmoil of Reconstruction. Southwestern Virginia drew its share of attention thanks to abundant mineral resources. The salt, lead, and iron works at locations like Saltville and Wytheville had stimulated some antebellum regional development, but a new wave of postwar speculators would attempt to expand regional investment. Confederate army cartographer Jedediah Hotchkiss (made famous as a mapmaker for generals Stonewall Jackson and Robert E. Lee) had noted exposed coal seams in southwestern Virginia and southern West Virginia during his Civil War campaigning. Based on these observations, he believed the Roanoke region could one day rival Pittsburgh as a center of American iron production, if only a rich enough iron ore deposit existed nearby.[8] Local boosters hoped Hotchkiss's prediction would prove true. A *Lynchburg News* correspondent, in an article that would attract regional and national attention as a reprint in the *American Farmer* and *Southern Planter and Farmer*, declared southwestern Virginia to be a landscape of "varied mineral deposits" that remained "literally a *terra incognita* to enterprise and capital." Here was a place, he continued, where "the great need is capital and enterprise."[9]

Likewise, Edward King appreciated what might be done locally if capital and nature were united. King was a roving reporter for the popular magazine *Scribner's Monthly*, and he journeyed throughout the South in 1873 and 1874, providing opinions on northerners' pressing questions: How was the South recovering from the war? Was it still an exotic region? Where are the best places to invest capital? King passed through the Roanoke Valley on his trip and saw in its farmland, forests, and mineral deposits great potential for a fusion of agriculture and industry. He asserted that the place "has a grand future. As a field for immigrants who have capital and intelligence, for the better class of large farmers, and for workers of metal, it cannot be surpassed."[10] Charles Rufus Boyd, a mining engineer, likewise touted the mineral wealth of Virginia's southwestern counties during the late 1870s and early 1880s, paying particular attention to both ores and the state of transportation and predicting big things for the future.[11] For all these expansive claims, the landscape King

FIGURE 5.1. One of Jedediah Hotchkiss's maps of central Appalachia, part of his work promoting the potential resources of the region. Jedediah Hotchkiss and D. C. Humphreys, *Map of Appalachian Virginia and Parts of W. Va. & Ky* (1873). Courtesy of Library of Congress.

rode through was a patchwork of tobacco, wheat, and corn farms, sprinkled with a few smelters, foundries, and saltworks, and enlivened by the occasional mineral springs resort.

Although Hotchkiss, King, Boyd, and similar boosters were decidedly optimists (and hardly alone in promising that some particular corner of the South was about to be the next big thing), portions of their vision for southwestern Virginia—and the Roanoke Valley in particular—proved quite prescient. King foresaw the expansion of the existing Atlantic, Mississippi, and Ohio Railroad (connecting Lynchburg to Bristol) via branch lines penetrating adjacent valleys, drawing natural resources out of the mountains and channeling them to cities. This would be key to the growth of Salem and Big Lick, and the Atlantic, Mississippi, and Ohio would form the core of the Norfolk and Western Railroad (N&W). King made a clear prediction for *Scribner's* readers: "As soon as the railway now prompting the growth of these interests can shoot

The Roanoke Valley, Virginia. [Page 577.]

FIGURE 5.2. The Roanoke Valley remained a sleepy farm community in the early 1870s. Illustration from a sketch by J. Wells Champney, in Edward King, *The Great South* (Hartford, Conn.: American Publishing, 1875).

out its feeders on either side, the number of tons of minerals annually exported from Virginia will be quadrupled."[12]

King was right. The railroad promoted mineral extraction from east-central Appalachia, and Roanoke experienced an incredible boom during the 1880s. One mineral in particular, coal, sat at the center of this natural wealth. Roanoke boosters touted the city's greatest natural advantage as "an ample and inexhaustible supply of the best coal at a minimum cost."[13] A local editor expounded on this wealth, envisioning the riches that might flow from a region where the "great seams of the 'Kanawha coal field' [exist] in the highest excellence of quality and quantity, underlying no less than 2,000 square miles, the seams in which would measure a quantity equal to 40,000,000 tons, enough to supply this country for a few hundred years." Indeed, he was amazed that this resource had gone so long without development, having "lain useless and profitless through all these busy years of railroad and iron-making progress. It is a happy reflection that this condition of things is now about to end."[14] Coal would drive regional economic development just as efficiently as it powered the Norfolk and Western's locomotives.[15]

Coal was the most important but not the only mineral located within convenient distance of the new commercial center. The New River Valley, southwest of Roanoke, had been home to companies mining and smelting iron, copper, and lead since the eighteenth century. King carefully noted that the region was thick with mineral resources: he listed coal seams, copper and iron deposits, salt, limestone, and lead as present in abundance. By the 1880s, zinc production joined the valley's industrial enterprises, and regional iron was of increasing importance in the 1800s. The particular qualities of New River iron ore and local forging techniques produced an iron especially well suited to making railroad car wheels, items in high demand as tracks spread across the rapidly expanding nation. Along with coal, metals mining encouraged Roanoke's growth as an industrial hub.[16]

Roanoke was not anomalous. For all the conceptions of Appalachia as a rural region, one of its essential characteristics during the last decades of the nineteenth century was the birth and growth of commercial towns. Although the region would never urbanize to the extent of New England or the mid-Atlantic coast, a wave of mountain towns sprang up after the Civil War, and Roanoke was but an extreme example. An increasing national hunger for coal and timber—especially for fueling new railroad lines and a construction industry turning to balloon framing—generated a need for company and service towns across the southern mountains. Established cities such as Chattanooga, Asheville, and Knoxville gained thousands of residents. Johnson City, Tennessee, an emerging iron and railroad center, increased in population rapidly between 1880 and 1890.[17] Famously, Birmingham, Alabama, became a significant coal and iron city in Appalachia's southwestern foothills. Perhaps more impressively, small crossroads, junctions, and sleepy farm towns like Big Lick bloomed into significant population centers in a span of just a few years or even a few months. Among these new entrepôts were places like Bluefield, Huntington, and Hinton, West Virginia; Pocahontas, Virginia; Anniston, Alabama; and Andrews, North Carolina.[18]

Two examples in addition to Roanoke stand out for their self-conscious boosterism and the importance of mountain environments for city creation. Middlesborough, Kentucky—now Middlesboro—epitomized the boomtown fever that accompanied these growing cities. Much as was the case with Roanoke, local boosters and foreign speculators had selected a site near the Cumberland Gap where Kentucky, Tennessee, and Virginia meet as a logical location for an important Appalachian industrial town in the late 1880s, and they set about transforming their dreams into reality. Alexander Arthur, a Scot, raised funds from English investors to establish Middlesborough, and he commissioned a promotional railcar that carried samples of Kentucky coal, wood, and iron ore throughout New England in search of investors. In an effort to sell the site's natural advantages, at each stop a magic lantern show aided in the pitch. Despite the development fever surrounding the town, it never became the industrial center Arthur envisioned, largely because the Cumberland Gap gradually faded as a transportation nexus in the rail age. Its geographical position and topography, which had once made the Gap so important, were less crucial by the late nineteenth century.[19]

The speculators who laid out Harriman, Tennessee, were just as ambitious. A land company backed by northern capital and supported by famed temperance advocate Clinton Fisk envisioned a prohibitionist industrial city in Roane County that would house steel, lumber, and textile factories. There too investors looked to abundant local timber, coal, and iron ore—former Union general John T. Wilder's Roane Iron Company was located nearby—to ensure a profitable industrial center. The land company engrossed more than half a million acres of East Tennessee countryside, and its initial land auction in 1890 attracted five thousand investors. Although speculators would pour more than $7 million into the Harriman venture, the economic crisis of 1893 curtailed the development of a city of half a million people that Fisk and his associates envisioned, and Harriman never became a significant urban center.[20] For every Middlesborough and Harriman, there were several more successful cities, and the wave of mountain boomtowns continued in the first years of the twentieth century. For example, the Tri-Cities of Bristol, Virginia, and Johnson City and Kingsport, Tennessee roughly doubled in size between 1900 and 1910.[21]

Roanoke rode the edge of this development on a language rich in environmental determinism. The *New York Herald* characterized the speculative energy itself as a force of nature: "From Roanoke, through Southwest Virginia, to Birmingham, Alabama, a wave of speculation is rolling, white capped with the dollars of the rich and the poor."[22] This urban expansion was not exclusive to Appalachia, of course. The growth figures for many of the era's principal southern cities were impressive. Between 1880 and 1890, Richmond increased its population from 63,600 to 81,338, Nashville from 43,350 to 76,168, Atlanta from 37,409 to 65,533, and Memphis from 33,592 to 64,495. During rapid urban growth across the mountains and the South more broadly during the 1880s, Roanoke's meteoric rise attracted its share of attention. It leaped from a hamlet of four hundred people to a city of twenty-five thousand during the decade. The new city seemed primed to enter the ranks of the region's premier commercial centers. Norfolk was the South's tenth largest city in 1890, with a population of just 34,871, a figure Roanoke was rapidly approaching.[23] Indeed, in less than ten years the Roanoke Valley went from a sleepy agricultural crossroads to hosting the state Democratic convention, with the bustling proceedings recorded for a national audience in the *New York Times*.[24] Within and without the city, people were coming to imagine Roanoke, nestled in its valley so close to rich coalfields, as destined to be a great metropolis.

New buildings, even new neighborhoods, seemed to appear overnight. Novelist Charles Dudley Warner—who with Mark Twain coined the term

"Gilded Age"—visited Roanoke mid-boom and came away astonished by its rapid transformation. Warner found in the boomtown the industrial energy and chaos of the era, and he saw a place where the gilding had been laid on in rather slapdash fashion. He wrote of "the noise of hammering and hauling [that] filled the air; streets of temporary wooden shops and dwellings, drinking shops, and 'hotels' with false board fronts hiding the upper half stories, and big-letter signs, after the manner of the West, isolated dwellings on every hill and knoll, everywhere the debris of building and ditching and road making."[25] The built environment and the environment of building overlapped in a frenzy of development.

It is revealing that Warner drew on the American West as a reference, since the similarities between Appalachian and western boomtowns were substantial. The rapid construction of western cities such as San Francisco, Sacramento, Denver, and Pueblo also often drew on the wealth-generating potential of mineral extraction and the seeming abundance of rural landscapes, mated to the motive power of the railroad.[26] Such was also the case for the hub of the "Great West," Chicago, following the Civil War.[27] In both the post–Civil War West and Appalachia, the idea of wild (or at least under-used) land open to the transformative power of capital was powerful. Western cities and the extractive industries that fueled their growth promised to transform "wilderness" into productive territory, to transform nature into culture, and, on a smaller scale, industrial cities like Roanoke seemed to offer the same for the southern mountains, long envisioned as a relatively marginal space bypassed by the earlier waves of American development.

The significance of the explosive growth that took place in Roanoke in a single year—from 1881 to 1882—cannot be overemphasized. The town grew from 58 houses to 268 in just twelve months. In addition to these private homes, a spot on the map that had possessed little in the way of manufacturing, business, or entertainment now held nine hotels, five restaurants, twelve boardinghouses, five tobacco factories, four shoemakers, seven blacksmiths, three stove makers, two jewelers, a foundry, two planing mills, two wheelwrights, one wagon spoke factory, two agricultural implement manufacturers, one cigar factory, twelve saloons, and a bowling alley. There were two newspaper offices competing to trumpet Roanoke's newfound glory. Adding to the bustle and din, a stockyard established on the edge of town the same year held hundreds of milling cattle and sheep on their way from mountain ranges to eastern tables.[28] The first issue of one of Roanoke's new newspapers, the *Roanoke Leader*, gave evidence of this frenzy of development, touting the new community's seemingly limitless prospects. Its pages were crowded with ads for construction companies and 2,500 town lots for sale, and the lead article asserted that the city was planted in a beautiful and healthy site, "admirably adapted by nature for the location of a city," located as it was in a fertile valley ringed by mountains that seemed destined to funnel future rail lines.[29]

More than anything else, it was the expansion of the Norfolk and Western Railroad that had boosters excited. Indeed, the railroad all but created Roanoke. Before the Civil War, Virginia had invested more energy and money into its railroads than any other southern state, and after the war's ruinous conclusion for the Confederacy, the commonwealth turned again to railroads to foster economic prosperity. During the late 1860s the legislature took drastic action, ceding control of state interests in various railways to private industrialists in the hopes that they could rebuild and then expand the lines and stimulate recovery. As an additional inducement, Virginia would remain on the hook for all existing railroad debts. Californian Collis P. Huntington, a co-owner of the Central Pacific, the western branch of the first U.S. transcontinental railroad, acquired the Chesapeake and Ohio, which would connect eastern Virginia's ports to the Ohio River and beyond. Railroad executive and former Confederate general William Mahone in turn received the rail network that would become the N&W, and he would base the new leviathan's corporate headquarters in Roanoke. The Chesapeake and Ohio and the N&W would both be important players in state politics as well as the economy, purchasing Richmond newspapers and bankrolling favored politicians.[30]

Norfolk and Western's Roanoke operations quickly dwarfed all other city businesses. Between them, the Roanoke Machine Works, the Roanoke Railroad Yards, the railway offices, and the affiliated Shenandoah Valley Railroad employed almost one thousand workers.[31] City land near the tracks attracted ancillary businesses like machine shops and metal works and other operations, such as a hoe handle shop, that subsisted on the timber brought in by the trains.[32] The stockyard that sprang up also relied on the cattle and hogs of the New River Valley carried into the city on the railcars.[33] The *Roanoke Leader* captured the influence of the railroads in an 1883 article that walked readers through the new Roanoke Machine Works, marveling at its size and complexity. It was a facility with an engine house with "stalls for twenty engines," where power for the facility's tools and machinery came from a one-hundred-horsepower steam engine and a boiler "seven feet [in] diameter and twenty-eight feet in length." The complex also housed shops for railcar construction, wheel making, lumber drying, coppersmithing, blacksmithing, and all the attendant enterprises requisite to keeping a railroad running. Ultimately, out of frustration with his inability to adequately represent the size and complexity of the shop, the editor advised Roanokers to see it for themselves: "Nothing less than a day would do" to tour the works with any thoroughness.[34] For the editor, the Roanoke Machine Works encapsulated the power and influence of the railroads and coalfields that were the city's raison d'etre.

Roanoke's booming growth quickly spread to nearby communities. Salem, the quiet valley village that predated Roanoke, became a satellite town of the new hub. One enthusiastic supporter looked askance at Roanoke's rise, claiming that Salem held "equal advantages for iron manufacture and railway operations." By 1890 its population had grown to roughly four thousand residents. And Salem too relied on industry associated with the railroad and urban construction, with a roster of business that included an iron furnace, a textile mill, multiple brick and lumber works, a wagon manufacturer, and machine shops. A proposed steel mill was even in the works.[35]

Rural districts deeper in the mountains were also connected to Roanoke's urban growth, tied to the burgeoning city by the rail lines that were stretching like steel fingers up regional valleys to capture timber, coal, and iron ore. Wherever the rails passed, growth followed soon after. Or perhaps it would be more appropriate to say that wherever natural resources lay, the railroad companies sought them out, spending money and bringing development as they went. Following the Civil War, as one scholar notes, "the railroad meant success or failure for communities of all sizes."[36] (And other Appalachian regions also turned to railroads to promote economic growth and sectional power, as did Asheville, which hitched its fortunes to the tourism and industrial development that civic leaders hoped would come with completion of the Western North Carolina Railroad.)[37] Roanoke and its hinterland towns and rail depots comprised a system as much as they were discrete places. Symbolizing the centrality of these railroads to the region, by 1884 the last page of the *Roanoke Leader* was dominated by timetables and connections for the railroads that made their way through the city. Their names—the Richmond and Allegheny, Norfolk and Western, Shenandoah Valley, Richmond and Danville, and Virginia Midland—formed a roster of regional enterprise and highlighted Roanoke's crucial links to both the rural mountains to the west and eastern industrial centers.[38]

The hinterland most crucial to Roanoke's rise, the place where the most important rails ran, was the vast coal district of southwestern Virginia and southern West Virginia then under development: the Flat Top and Pocahontas fields. Hotchkiss's notice of extensive coal seams in McDowell County, West Virginia, and Tazewell County, Virginia, during wartime campaigning had proven a catalyst for regional growth. After the war's end, Hotchkiss envisioned tapping these coal resources to transform western Virginia's small iron foundries into an industrial concentration to rival Pittsburgh. To this end he commissioned another former Confederate officer, Isaiah Welch, to survey the

Flat Top Mountain region and the tributaries of the Tug River in the spring and summer of 1873. Welch returned from his trip with reports of astounding quantities of high-grade bituminous coal visible in surface outcrops on the region's slopes. Hotchkiss, Welch, and other speculators bought land throughout the Pocahontas field and attracted interest from Frederick Kimball, at the time the Norfolk and Western's vice president, who secured funding from Philadelphia investors to extend a rail line into McDowell and Tazewell in order to bring the coal to market.[39]

It was the intersection of this abundant coal with the region's other valuable minerals that was especially exciting to railroad men and manufacturers. Kimball, who was also president of the Shenandoah Valley Railroad, along with Norfolk and Western president George Tyler, touted the landscape through which their railroads passed as "known to be rich in minerals, in fact supposed to be richer than any other portion of the Union."[40] A chemist hired to assess investment prospects adjacent to the new lines made a virtual guarantee of "a handsome profit for capital judiciously managed."[41] A similar study commissioned by the Norfolk and Western declared that because of "the occurrence here of a first-class and cheaply mined iron ore; the proximity of a magnificent coking coal-field; with limestones for fluxing purposes everywhere throughout the region; with a constant supply of pure water; surrounded by a fertile agricultural and grazing district capable of supporting a large population; and with numerous eligible sites for manufacturing purposes, this New River–Cripple Creek region certainly offers unusual advantages for the investment of capital."[42] Geology seemed to reflect the shining face of fate, or so said the railroad men.

A branch of the Norfolk and Western pierced the mountains and reached the rich Flat Top district of the coalfields in 1883. Along the way the railroad encouraged new towns to pop up like mushrooms after a rain. During the 1880s Tazewell County, located along the border of Virginia and West Virginia and in the coalfields, saw its population grow more than 50 percent, from 12,861 to 19,899. By 1888, a tunnel through Flat Top Mountain connected Roanoke's feeder lines to McDowell County, West Virginia, where the population expanded from just 3,074 people in 1880 to 7,300 in 1890 and 18,747 by the turn of the century.[43] Coal production swelled even faster than the mining population. McDowell grew in just twenty years from a relatively minor source of coal to the center of national production. In 1889 the county produced 246,000 tons of bituminous coal; in 1910 its miners dug twelve millions tons, most of which passed through Roanoke.[44] The tracks sought out the metal mining and smelting enterprises of the New River Valley as well as the coal seams, extending "like tendrils along the New River, up Cripple Creek, up

Little Reed Island Creek, and finally up Chestnut Creek."[45] Roanoke's boom was thus inextricably connected to rural growth in nodes along its railways, connecting town and country through a relationship to the mountains' natural resources. As Kate Brown notes of the growth of industrial cities during this era, "The urban grid was a concentration of the expanding rural grid, which linked the hinterland economically and spatially with cities. As a consequence, there were no topographical limits to urban space, and the cities grew and multiplied."[46] Tazewell, McDowell, and the Cripple Creek smelters could no more boom without Roanoke than Roanoke could flourish without its hinterlands.

Regional growth was important beyond local concerns, because it stoked development in more distant regions too. Roanoke's coal mattered a great deal to a nation in the throes of rapid industrialization. The Pocahontas field and Roanoke's rail hub were carefully watched by business interests as far afield as New York, where the *Times* assured its readers in 1902 that local labor unrest was temporary and that Roanoke's "coal fields [were] open," ready to supply northeastern factories and homes.[47] The Deep South cotton belt and the Piedmont tobacco districts also increasingly relied on Roanoke's rail network. In 1891 the Roanoke and Southern Railroad connected the city to Winston-Salem, North Carolina, a growing textile and tobacco manufacturing city, and from there to points south. One of the most important commodities hauled over the new line was sulfuric acid, manufactured from the high-sulfur iron ores of the New River Valley. Sulfuric acid was vital to the agricultural South because it was needed to manufacture eastern Carolina phosphate rock into fertilizer, increasingly important on the tired old plantation lands throughout the South.[48] From chilly sitting rooms in Boston to the flat cotton fields of South Georgia, Roanoke and its commerce were vitally important.

Roanoke's visionaries believed that an even brighter future lay ahead in the twentieth century and were supremely confident in the power of natural advantages to promote industrial development and economic success. They touted that a triumvirate of coal, iron ore, and cheap labor would make Roanoke the industrial heart of America. This rhetoric appeared in descriptions of numerous city endeavors. A steelworks proposed for the valley promised to forge a new Pittsburgh by taking "the crude material from the everlasting hills and [manufacturing] it into useful forms . . . for an annual sum of fully fifteen million dollars."[49] A single issue of the city newspaper in 1884 outlined plans to recruit a heavy gun foundry, a new textile mill, and a state normal school that would employ local resources and educate local people.[50] Another article expounded upon Roanoke's advantages over northern industrial cities; southwestern Virginia enjoyed "cheaper raw material, cheaper labor, and a more favorable climate" than its competitors. The author positioned Roanoke

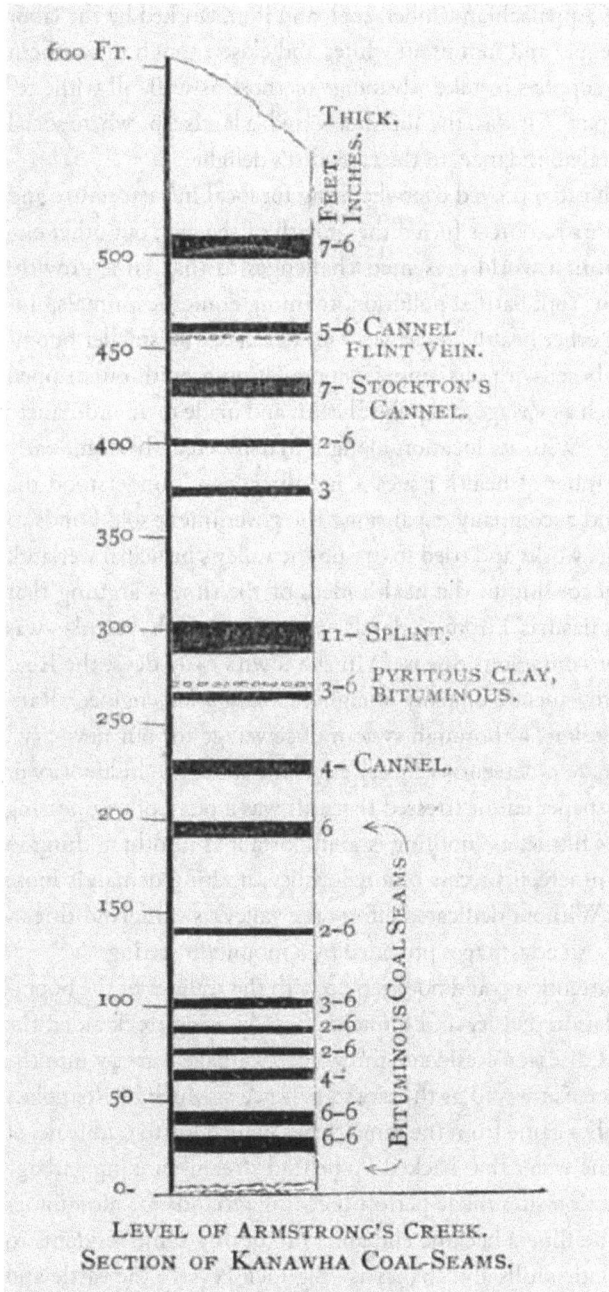

600 FT.

THICK.

FEET.
INCHES.

500 — 7–6

5–6 CANNEL
450 — FLINT VEIN.

7– STOCKTON'S
CANNEL.

400 — 2–6

3

350 —

300 — 11– SPLINT.

3–6 PYRITOUS CLAY,
BITUMINOUS.

250 —

4– CANNEL.

200 — 6

150 —

2–6

100 — 3–6
2–6
2–6
4–
50 — 6–6
6–6

0–

BITUMINOUS COAL SEAMS

LEVEL OF ARMSTRONG'S CREEK.
SECTION OF KANAWHA COAL-SEAMS.

FIGURE 5.3. Coal seams visible at one location along the Kanawha
River in West Virginia, indicative of the region's mineral wealth.
Illustration from a sketch by J. Wells Champney, in Edward King,
The Great South (Hartford, Conn.: American Publishing, 1875).

at the intersection of Appalachian timber, coal, and iron, worked by the labor of Piedmont freedpeople and mountain whites and close enough to southern cotton and tobacco supplies to take advantage of those as well, all while recruiting northern capital.[51] It was, the author asserted, a landscape where social inequality met natural abundance, to the capitalist's delight.

Roanoke's growth often proved overwhelming for local infrastructure and environments. Nature's resources fueled the growth of the city, but other elements of the nonhuman world presented challenges to that same growth. Major cities like New York battled pollution, roaming domestic animals, unsanitary streets, and other health "nuisances" for decades, and smaller boomtowns were equally beset with challenges as population growth outstripped capacities in areas such as sewage and public health and made town ordinances outdated overnight.[52] With its location along a marshy creek bottom, early Roanoke faced a number of health issues. City officials well understood the challenges that would accompany expansion. The government sold bonds to fund various drainage works and tried to dry up the valley's principal wetland, Long Lick, which according to the health ideas of the time—arguing that contaminated air ("miasmas") from wetlands was a grave health hazard—was a potential source of contagious disease.[53] In the town's early days, the Roanoke Land and Improvement Company brought in a sanitation engineer, Randolph Herring, to design "a thorough system of sewerage for our new city." Highlighting legitimate nineteenth-century concerns about the insalubrity of city life, a local newspaper editor stressed that this was a piece of engineering crucial for Roanoke's future, as "nothing is so important as health; nothing so necessary to the commercial success of any locality; nothing demands more careful attention."[54] Without dedicated efforts the valley's swampland threatened to undo the health advantages provided by a mountain setting.

But even careful attention could not keep up with the realities of the boomtown. A high population density and limited, crude housing stock aided the spread of contagious diseases. Residents threw their garbage directly into the streets and existing creeks served as the city's first sewers. Much of Roanoke's drinking water supplies came from the same creeks, which led to outbreaks of typhoid fever, and the water that backed up behind stream-choking garbage and around construction sites made perfect breeding grounds for mosquitoes carrying malaria. The illness became endemic, prompting some residents to refer to its symptomatic chills and shakes as "Big Lick Fever." The cattle and hogs that moved through the city's stockyards also likely contributed to the population's ill health. The city even suffered a smallpox scare in January 1883, but it escaped an outright epidemic.[55] As one local historian recorded of the city's early years, "Families in many instances are greatly crowded for comfort and perhaps for health . . . persons who occupy second stories are accustomed to throwing the accumulations of both chamber and kitchen into their yards, while others dig holes in their yards and there deposit filth that causes stench that is offensive to the olfactories of the neighborhood and it is inimical to health."[56] Even a decade after Roanoke's initial population explosion, the urban sewer system had not quite caught up. A severe thunderstorm in August 1892 caused flooding in the valley's lowlands, due in part to an incomplete drainage network. The flood resulted in more than $100,000 damage and the death of a Roanoke Machine Works blacksmith, who fell into a flooded construction site and drowned.[57] Although located in a stereotypically rural region, Roanoke replicated many of the contemporary environmental problems of the growing cities across the United States.

While urban residents commented on the ways in which the valley's environment changed so rapidly, these transformations did not entirely displace the older rural landscape. The creeks that served as sewers, and occasionally backed up to the distress of those living along them, continued to move water through the valley's wetlands, and the hills surrounding the town remained largely forested during the city's first decades, serving as places of retreat for residents weary of the booming rail center. One city historian recorded the plentiful fish and game still to be found within the city limits or nearby throughout the 1880s, noting that "pheasant (ruffed grouse) were found and shot on Mill Mountain; beaver were trapped on the Roanoke River; bass, trout, carp, sun perch, cat fish were teeming in every stream."[58] Roanoke remained very much a part of the broader mountain environment through more than just its exploitation of natural resources.

Mountain industry and wildlife could intersect in interesting ways, as outsiders attracted to industrial work in Roanoke also delighted in the sporting opportunities that remained on the edge of town. Sometimes these relationships highlighted just how quickly the rise of the city had changed the ecology of the Roanoke Valley. For example, T. J. Houston, a Pennsylvanian who had moved to Roanoke to manage the Crozer Steel and Iron Company's Roanoke operations, invited a northern friend on a hunting expedition in the valley in 1889. Houston's work kept him from hunting, but his friend (who remained anonymous in a magazine article recounting the trip) roamed the mountains around the city with an African American guide named Moses Johnson. They heard from a local farmer "that pheasants and wild turkeys were found up a mountain ravine in close proximity to us," and they saw a woodcock in the hills. But ultimately they discovered that Roanoke's population had largely stripped the adjoining mountains of woodland game. The transformations had, however, created an abundance of new edge habitat on the valley floor that supported certain game species. In the pastures and farmland there they found coveys of quail and took a nice bag of game birds before catching a train back to town, riding the rails that had fostered the transformations.[59]

Roanoke's rapid growth held the potential to disrupt social as well as material environments. As in other industrial towns in the post-Reconstruction South, race and class were intellectual concepts used to order space, and they were points of contention. Many well-off Roanoke residents were decidedly pro-growth, but at the same time they worried about the changing demographics of their city and the fact that much of the money and expertise that fueled expansion came from northerners.[60] Roanoke by 1890 had a substantial black population who mainly worked at service jobs and in manual labor. African Americans made up 30 percent of the city's residents but lived in segregated communities that remained largely hidden from the white view.[61] When blacks and whites did interact, it was often through formal relationships that stressed social differences, as when Houston's friend came to Roanoke to hunt and relied on the services and labor of Moses Johnson as hunting guide. (Highlighting the social distinctions that adhered to each role, the Pennsylvanian wrote of what he saw as physiological and psychological differences between himself and Johnson, describing the black man in animalistic terms as "a dark mulatto of good physique, surpassing foxhounds in endurance.")[62] Another substantial portion of the city's population comprised white farm residents just arrived from the surrounding mountain counties, drawn to Roanoke by a profusion of new industrial jobs. This was not uncommon in the postwar South. Referring to southern cities and their reliance on rural migrants for growth during this period, one historian notes that "it was the people above all who gave the southern city its rural atmosphere."[63]

These same industrial forces reshaping people's use of nature directed migration across the other stretches of the transportation network that connected Roanoke to the hinterlands too. As one early branch of the Great Migration, African Americans from the plantation South moved into the mountains and coal mining in large numbers seeking greater economic opportunity. The number of black miners in Tazewell and McDowell Counties swelled, and by the early twentieth century roughly one in five regional coal miners was African American. Rural mountain whites also migrated to find jobs in the coal mines and the cotton mills springing up in the company towns lining the southern rail system in both the mountains and the adjoining Piedmont, fast becoming the nation's new textile hub. Roanoke itself was soon home to a substantial spinning industry.[64]

Roanoke's "rural atmosphere" did not always equate to tranquility, of course. This diverse, rapidly growing population occasionally came into violent conflict. In 1892, William Lavender, a local African American man, was accused of the attempted rape of a twelve-year-old white girl. An angry white mob captured Lavender and engaged in a form of public vigilante action growing toward epidemic scale in the last decade of the nineteenth-century South: they lynched him. At the gruesome scene a reporter for the *Roanoke Times* recorded that Lavender's body was left "dangling by a rope an inch in diameter from the limb of an oak tree on the bank of the Roanoke River . . . frozen stiff and stark."[65] Another spectacle lynching followed the next year. The beating and robbing of a white woman in the city touched off a riot that left eight Roanokers dead and thirty more injured. Culminating the violence, on September 21, 1893, a mob seized Thomas Smith, an African American man, and even though the victim failed to positively identify him, the enraged crowd hanged Smith. Pieces of Smith's clothing and bark from the gallows tree would circulate the city afterward as souvenirs. The lynching party then took his body to the house of Roanoke mayor Henry Trout—whom many white residents accused of encouraging city police to use violence to suppress white rioters—with the intention of burying Smith in the mayor's yard. Instead the mob built a bonfire and burned the corpse in a horrific public scene that drew as many as four thousand spectators.[66] Roanoke's violence was not exceptional in the contemporary South—the New River Valley mining district also saw its share of lynchings around the turn of century—but the city's riots drew state and national attention and widespread condemnation.[67]

The deaths of Lavender and Smith, although part of a broader trend of spectacle lynchings across the Jim Crow South, also highlighted the tensions created by Roanoke's meteoric growth and the physical and social distance that remained between black and white residents. The lynchings highlighted the new landscape of southwestern Virginia and Roanoke: the burgeoning city was a product of spatial relations and changed them at the same time. Its industries and corporate leaders built tracks into mountain valleys and coalfields, setting enormous quantities of coal into motion and spawning numerous ancillary businesses in Roanoke and other portions of Appalachia. Northern speculators invested streams of capital in southwestern Virginia and southern West Virginia. People moved with these flows too. Fleeing poor prospects on farms or attracted by the prospect of steady wage work, job seekers flooded into Roanoke to work in railroad shops and warehouses, while others moved to Tazewell or McDowell Counties to mine coal. These flows changed the physical and social environments of central Appalachia in ways that were not always attractive.

By the early decades of the twentieth century Roanoke was a place in the mountains, supremely reliant on mountain resources but set apart. At the turn of the century other industries had joined the city's railroad and coal businesses: the Norwich Lock Manufacturing Company, Cushman Iron Company, West Roanoke Iron Company, a brick factory, Crystal Spring Soap Factory, and the Bridgwater Carriage Company all contributed to Roanoke's economy, helping it weather the economic downturn in 1896. The financial troubles of the 1890s were short-lived for the region; between 1900 and 1910 the city continued to grow, adding more than seventeen thousand residents.[68] Its citizens enjoyed Roanoke's modern conveniences and industrial might. They took pride in a public school system rivaling those of Richmond, Petersburg, and Norfolk; the rayon manufacturer American Viscose Corporation, which employed thousands of workers in a cutting-edge industry; attractive streetcars; paved roads; electric lighting; a country club; and Roanoke College, founded within a decade of the city's boom and complete with a laboratory for a "Department of Chemistry and Physics," where a professor with "three years' study in German universities" taught students from as far afield as Japan, Mexico, and "Indian Territory." Dwarfing all these enterprises was the vast complex of the Norfolk and Western, with its numerous machine shops, warehouses, and now the South's largest locomotive plant.[69] As dependent as Roanoke was on mountain coal, it seemed a different world than the impoverished coalfields to its west.

Perhaps nothing better emphasized the physical connection and the intellectual distance between modern Roanoke and the coal districts on which it relied than the events of one August day in 1921. That afternoon seventeen aircraft under the command of Major Davenport Johnson landed at Roanoke's new airstrip, where the modern machines attracted a great deal of public attention. The mayor met the aviators at the airfield, the pilots stayed at the luxurious Hotel Roanoke (a mock Tudor pile downtown), and American Legion volunteers guarded their planes that night. A large crowd of city residents assembled the next day to see the planes off, bringing business around the airfield to a near standstill. The squadron was not part of a barnstorming tour. It was on its way to West Virginia, where its pilots would become part of the violent government-endorsed effort to disrupt coal miners' attempts to unionize, in a conflict alternately dubbed the West Virginia Mine War and the Battle of Blair Mountain. As trains of bituminous coal rolled into Roanoke from the west, bringing the wealth of West Virginia's coalfields through Roanoke and to the broader world beyond, the planes flew toward the mines, where their

FIGURE 5.4. Early-twentieth-century Roanoke housed industries reliant on rural people who had recently moved to the city, including textile mills like the one that employed these children. Photo by Lewis Hine, 1911. Courtesy of the Library of Congress.

pilots would work to ensure that the trains continued to run and that the coal that filled them was as cheap as possible.[70]

Highlighting Roanoke's growing psychological separation from its mountain resource base was its residents' rising distaste for the city's "Appalachian" label. When a government planning agency, the Appalachian Regional Commission (ARC), began directing federal funding flowing into the mountains in the 1960s, Roanoke County was not one of the official ARC counties, despite its location well within the mountains and its economic ties to the coal belt. (This was the case despite the fact that the ARC's liberal definition of Appalachia included counties as far afield as southwestern Ohio and northeastern Mississippi.)[71] Roanoke's exclusion from the new federal organization came not from an outside slight but from local desires. Many Roanoke Valley residents, especially civic boosters, resisted the "Appalachian" label, fearing that connection to the War on Poverty would bring with it "hillbilly" stereotypes and negative press. Several other Virginia counties reacted the same way, and the Shenandoah Valley also remained outside of the ARC boundaries, thanks in large part to Richard Poff, a local congressman who pushed back "against government activism" in the lives of mountain people.[72] Roanoke residents proclaimed themselves residents of modern industrial America, enjoying the newest advances and conveniences of a mighty nation, even if that might came

FIGURE 5.5. Downtown Roanoke, showing the rail lines that bisected the city. Photo by W. Carlton Parker, 1931. Courtesy of the Library of Congress.

in part from their city's place in the mountains and its exploitation of their natural resources.[73]

What then was unusual about Roanoke when it comes to the Appalachian experience? Very little, really. A city within a region characterized by its rural nature seems distinctive, but of course a significant number of Appalachian people lived in towns or small cities, including Asheville, Charleston, Knoxville, Chattanooga, Huntington, Bristol, Kingsport, Johnson City, and smaller centers.[74] By the time Roanoke bloomed, larger cities also existed on the margins of the southern mountains—Birmingham, Atlanta, Charlotte, Pittsburgh, Cincinnati—serving as places of employment and commerce for Appalachian residents. For every sedentary resident of a forgotten hollow there was a mountaineer who lived in town, had relatives who did so, or visited with some regularity. Roanoke's explosive growth was exceptional, and few towns in Appalachia or elsewhere in the United States grew so large so quickly. But many other Appalachian cities emerged from natural resource booms that sent various commodities from the mountains to the lowlands, and those booms were connected to the development of railroads. Roanoke, like other Appalachian cities, emphasized the importance of connectivity and commoditization in the region. It was a place born of salt licks and settlers' traces, rails and crossroads. It boomed funneling mountain coal to eastern cities, and it filled up with northeastern money and rural southern workers. By 2010, Roanoke had grown to resemble other mid-sized American metropolitan areas, a landscape of approximately two hundred thousand people (Virginia's fourth largest urban concentration) characterized by the typical politics pitting the city's core against its suburban fringe. In some respects it had become Anywhere, USA.[75]

CHAPTER 6

Scenery

Recreation and Tourism
on Grandfather Mountain

IT WAS AN ODD GROUP that gathered at the new mountain resort of Linville, in the remote northwestern corner of North Carolina, for a Fourth of July celebration in 1892. One local in attendance declared the lowland socialites who surrounded him—who had come from as far away as Philadelphia as well as closer cities like Wilmington, North Carolina—"the finest gathering of high class people that ever greeted the opening season of a new mountain place in North Carolina."[1] These wealthy men and women crowded the porches of the brand-new Eseeola Lodge, a sprawling log hotel clad in rustic chestnut bark siding, intent on taking in a range of "traditional" Appalachian festivities "performed by the natives around Grandfather Mountain." The "cultural" exhibitions put on by mountain residents included sack races, a greased hog chase, and an ox race.[2] The wealthy MacRae family and Samuel T. Kelsey, the resort's developers, hoped the celebration kicking off its inaugural summer would lure some of the wealthy tourists into purchasing vacation homes and make Linville the next fashionable nature resort in the southern mountains.

The developers' aspirations drew on a long history. Wealthy Americans, particularly from the South and the mid-Atlantic coast, had long taken summer vacations in southern Appalachia. Lowland residents had flocked to the region's numerous hot springs—the most famous of which was White Sulphur Springs in modern West Virginia—believing that the waters contained healing properties and to rub shoulders with fashionable company. Likewise, towns like Greenville, South Carolina, and Asheville and Flat Rock, North Carolina, had attracted well-to-do planters escaping the heat of the Piedmont, Lowcountry, and Gulf Plain. These resorts also served as refuges for wealthy white families hiding themselves and their slaves from the main columns of the Union army during the Civil War. Newer communities, such as Highlands, North Carolina, joined the fold after the Civil War. Linville's promoters hoped they could take advantage of this entrenched model.[3]

FIGURE 6.1. Appalachian resorts like Linville often sought to combine outdoor recreation with stereotypical ideas about regional rusticity. "Sleighing—Linville, NC," Howard Pelton, 1912. Courtesy of the Library of Congress.

By the late nineteenth century, regional boosters like the MacRaes and Kelsey sold (and tourists consumed) a particular sort of commodity: scenery. A confluence of forces served to make Appalachia an increasingly attractive place for tourists after the Civil War. The nation's growing urban population, a burgeoning prosperous class of white-collar workers with newfound leisure time, expanding rail networks, and the "discovery" of the southern mountains as a distinctive cultural region by a host of writers all generated new interest in Appalachia. The aesthetic value of scenic places lay at the core of late Victorian- and Progressive-era nature tourism, replacing earlier fears of "wild" spaces with romantic ideas about the uplifting and spiritually fortifying qualities of mountain landscapes in particular. Appalachia—though less heralded than wilder and more geologically dramatic western landscapes—promised both physical and emotional pleasure. These conceptions of the worthy qualities of certain landscapes contributed first to regional tourism and then to effort to conserve or preserve particular sites.[4]

Contemporary ideas about health and the body were also connected to the benefits of scenic mountain locales. National figures like Theodore Roosevelt touted the physical and emotional benefits of a vigorous life in rural settings. Doctors diagnosed countless Americans as suffering from various nervous problems stemming from the debilitating brainwork of the cities and prescribed time outdoors in rustic settings like the mountains. And conceptions of the environmental roots of diseases like hay fever sent thousands on annual pilgrimages to the "fresh air" of the highlands.[5] Linville and the countryside surrounding Grandfather Mountain were perfectly positioned to take advantage of these forces, benefiting from these beliefs in the physiological and emotional value of beautiful scenery and its accompaniment, fresh air.

The story of Grandfather Mountain and Linville Resort also highlights another important aspect of emerging Appalachian recreational landscapes: the fusion of private and public development. Although historians have focused much more on the creation and management of public lands in the mountains, private development of various attractions was an important force in transforming a landscape many writers criticized as wild and uncultured into a pleasing rustic and scenic playground.[6] Linville epitomized this development. It was founded as an investment venture by capital from "Boston, New York, Philadelphia, North and South Carolina, Missouri and Kansas," and the MacRae family would also eventually develop nearby Grandfather Mountain as a tourist attraction.[7] Like many of these initially private sites, over time parts of Grandfather would become public. The Blue Ridge Parkway corridor, under the authority of the National Park Service, would cut across its eastern flank (the last link was completed in 1987), and North Carolina assumed control of the majority of the mountain as a state park in 2008. A nonprofit organization still maintains operational control of one end of the mountain, carrying on the site's long history of private management. Public-private cooperation would continue to be an important part of Appalachian development throughout the twentieth and early twenty-first centuries, characterizing the growth of resort communities such as Asheville, conservation organizations like the Nature Conservancy, and regional outdoor attractions like the Appalachian Trail.

સ્જ

Even in a locale as remote as Linville, which sits in modern-day Avery County, carved out of adjoining Watauga, Caldwell, and Mitchell Counties in 1911, scenic tourism predated the MacRaes' and Kelsey's efforts. Grandfather Mountain had long been an attraction for visiting scientists, especially the botanists discussed in chapter 2. André Michaux, François André Michaux, John Lyon, Elisha Mitchell, Moses Curtis, Asa Gray, Leo Lesquereaux, Amos A. Heller, Charles Sargent, and John Harshberger all visited the peak and surrounding valleys. These visitors spread word of the area's rugged terrain, scenic charms, and seemingly isolated people, as well as its plant diversity.[8] The region's bard arrived later in the nineteenth century, however. The most outspoken advocate for the tourism potential of the High Country (as the immediate region came to be called) was an eccentric writer, educator, and entrepreneur named Shepherd Dugger.

Unlike the botanists who preceded him and many of the numerous later promoters of the area, Dugger was a native of the mountains. He was born into modest circumstances in 1854 in nearby Johnson City, Tennessee, and his family moved to Banner Elk, North Carolina (about five miles from Grandfather Mountain) in his infancy when his father found employment in the local Cranberry iron mine and forge. After attending the University of North Carolina for a year in the early 1880s, Dugger married Margaret Calloway, whose family lived at the base of the mountain. With the backing of his new father-in-law, Dugger built a two-story inn called the Grandfather Hotel near the headwaters of the Watauga and Linville Rivers and set about promoting the natural splendor of the region. The hotel was a modest success. Dugger listed among his guests from 1885 to 1890 senators, artists, judges, amateur botanists, and the future president of the Chicago World's Fair, H. N. Higinbotham.[9]

Dugger portrayed the Grandfather Hotel as a place where guests would feel particularly in touch with nature, a restful establishment at the foot of a mountain thick with imposing virgin stands of eastern hemlock, red spruce, Fraser fir, and mountain ash.[10] A printed ad for the hotel billed it as "a well-kept house, in a most delightful spot, and watered by one of the best springs in the region."[11] The first year of business was more profitable than Dugger had expected. In an autobiographical manuscript he prepared later in life, Dugger reveled in the heady success of 1885. "All the people going to Grandfather Mountain, came to our place and traveled our path, passing the stone-face and the cold spring, to the top, which gave us a rousing business that summer."[12] An anonymous visitor, later writing to the national magazine *Christian Observer*, concurred that Dugger's hotel offered great entertainment in the form of pristine scenery. Of the sunset view on top of Grandfather, she or he wrote, "No mortal man would ever have guessed what was in store for us. The whole scene surpassed the range of man's imagination, and no artist's hand could ever paint it. We saw the glories of earthly beauty, and a slight manifestation of the shining portals of heaven."[13]

The advertisement for the Grandfather Hotel (and the broader region) that reached the largest audience was an 1892 novel written by Dugger, *The Balsam Groves of the Grandfather Mountain*. Describing an ascent of the mountain by a group of well-to-do tourists from North Carolina's lowlands, the novel's purple prose focuses on scenic nature as the area's greatest resource. Echoing testimony of real visitors to the region, the characters celebrate the mountain as an untouched spot "clothed with ferns, mosses, mitchella and oxyria, and

FIGURE 6.2. "Panorama West and Southwest from Grandfather Mountain. Typical Southern Appalachian Mountains," circa 1902. Photo in U.S. Department of Agriculture, *A Message from the President of the United States Transmitting a Report of the Secretary of Agriculture in Relation to the Forests, Rivers, and Mountains of the Southern Appalachian Region* (Washington D.C.: GPO, 1902).

supporting a mixed growth of red spruce (picea [*sic*] rubra) and balsam (Abies fraseri), whose matted branches form a beautiful green canopy."[14] For Dugger's characters, the unbridled nature of the place is intoxicating, as when "the jangling, twangling, tinkling chime of the bells, on the flowery, ferny brook-channeled slopes of the mountains . . . entranced them into a drowsy, soul-lulling stupor."[15] What more could weary urbanites desire?

Although the plot of the novel ostensibly revolves around a sentimental love story, its main purpose seems to be to transport the characters from one of the mountain's real scenic attractions to another. In an appendix to later editions of the book, Dugger made its promotional nature even clearer, noting the Grandfather Hotel as a comfortable and convenient base for exploring the region, furnishing good food, a healthy climate, and easy access to the trail that his fictional tourists use to scale the mountain. Swept away in expounding the restorative powers of mountain air, he rhapsodizes that the inn "nestles so near the evergreens that the sweet odor of the balsams is wafted in at the doors, and, sweeping through the commodious hallways, cures hay-fever and bronchitis, and prolongs the lives of consumptives."[16] Edward Bok, in a largely favorable review in *Ladies' Home Journal*, recognized the novel as fundamentally "an advertisement for that section of North Carolina in which Mr. Dugger has his habitation."[17]

As his claims about the curative properties of mountain air indicated, Dugger sold nature as both visual delight and medical treatment, and some guests bought the pitch. The same writer to the *Christian Observer* who so admired the scenery also testified to the salubrious effects of the Grandfather region, reveling in "the pure breezes of a healthful climate," which "cured one of the party of his habit of sneezing."[18] Dugger's claims that rural air and a favorable site promoted health were common assertions by tourism boosters and medical professionals in the late nineteenth century. Americans had a long history of believing that particular landscapes fostered either health or illness, and they connected bodily well-being to intrinsic qualities of the land. Soil types, prevailing breezes, vegetation, temperature, humidity, an abundance or absence of standing water: any or all of these environmental qualities were believed to affect the healthiness of a place. In particular, many people retained a belief that currents of damp or fetid air ("miasmas"), often associated with swamps or lowlands, carried sickness and disease, and they posited mountain air as the most healthful alternative.[19]

Health claims for rural areas were not restricted to southern Appalachia in the late nineteenth century. At the same time that tourists visited the Grandfather Hotel to recover their health, Americans in other regions were seeking salubrious places to vacation or relocate—the "country air" cure was widely regarded as a tonic. For example, resort hotels in the White Mountains of New Hampshire attracted Boston and New York City socialites seeking an escape from hay fever; Colorado's Front Range became a popular allergy refuge as the White Mountains grew more developed, only to be superseded by the dry air of mountainous Arizona and New Mexico; and from the mid-nineteenth century some consumptives found their cure in Southern California's coastal ranges. Chicago residents headed to Mackinac Island, at the juncture of Lake Michigan and Lake Huron, when city dust and summer heat triggered coughing and sneezing fits. Residents of flatter, less scenic locales made their own health pitches. Even the piney woods of the Florida panhandle and southwestern Georgia had their share of salubrious escapes.[20]

Closer to Dugger's enterprise, during the same era, other regional boosters touted the curative power of the southern highlands. Mountain mineral springs were popular destinations, from West Virginia's White Sulphur Springs to Tennessee's Red Boiling Springs.[21] South of Grandfather in the North Carolina mountains, Asheville had built a reputation as a healthy recreation destination in the years following the Civil War, as had the prestigious resort community of Highlands, developed by Samuel Kelsey during the 1870s. Northern-born, Chattanooga-based industrialist John T. Wilder, operator of the Cloudland Hotel on Roan Mountain (located west of Grandfather on the Tennessee and North Carolina line) also touted the healthy setting of his accommodations.[22] And located one valley north of Dugger's hotel, the Reverend Edgar Tufts, founder and head of a school that provided girls with religious and vocational instruction, sold "the invigorating climate, beautiful scenery and pure water" to prospective pupils.[23] The locale, Tufts promised, "offer[ed] to our students the most delightful and healthful climate to be found anywhere in the South."[24] The Grandfather Hotel was thus but a small part of a national and regional health-tourism craze.

These enterprises sprouted in a region that had previously been reliant on small-scale agriculture and what might be termed a "forest economy" of hunting, gathering, and timber cutting. Boosters sold a scenic and all-but-untouched nature, but human activity had significantly altered Grandfather Mountain's environs by the end of the nineteenth century. Recent Appalachian historiography has emphasized the extent to which the southern highlands were connected to outside regions through trade, agriculture, tourism, and industry during the late nineteenth century.[25] Appalachia had never been truly isolated, and the increased construction of highways and railroads in the late 1800s made travel throughout the region faster and easier.[26]

Residents of the three counties that came together at the highest point of Grandfather—Caldwell, Mitchell, and Watauga—relied primarily on agriculture for their living. In 1890, farms in the three counties were slightly smaller than the state average, and a higher percentage of their farmland remained unimproved woodlands rather than fields or pastures. This corner of Appalachia also lacked the important staple crops of the lowland South, with no cotton, and little tobacco outside of select valleys, and the value of farm implements and machinery was low. But numerous small farms did raise livestock and grow grain such as corn and oats.[27] Western North Carolina remained an open range for livestock grazing well into the twentieth century, and Dugger described cattle roaming the forests. While the rugged upper reaches of the mountains held little forage value, local farmers periodically assembled herds of cattle, hogs, and sheep on the lower slopes for drives to foothill towns like Morganton, North Carolina.[28] These livestock drives were a way to turn forest mast and excess corn into a saleable resource, one capable of walking itself to the market. Harvesting local plants such as ginseng to sell in surrounding towns was also a popular activity on the forested slopes.[29] Although few local residents were as strongly tied to outside markets as Dugger, who relied on tourists traveling on rail and carriage roads, most likely had some contact with regional market centers through the sale of surplus commodities. These commercial interactions provided much-needed cash for supplies, like coffee and sugar, not produced through home manufacture.

For all the importance of farming and open-range livestock, the local activity with the greatest impact on the environment was certainly cutting timber. Farmers and small-scale loggers had long cut Appalachian woods to clear farmland, furnish building materials, and earn cash, but mountain timbering accelerated after 1880, from the timbered ridges of West Virginia to the northern reaches of Alabama and Georgia. Between 1880 and 1900, logging in western North Carolina became an important industrial concern, though the companies involved tended at first to be relatively small-scale, usually local or state-level ventures. Most employed selective cutting, taking the most desirable species and leaving the rest, and they mainly employed human, animal, and water power. After 1900, however, large, well-capitalized corporations— most often based outside of the region—became the dominant actors, using narrow-gauge railroads and steam winches to access increasingly remote tracts of timber. Many of these operations relocated to the southern mountains from the pine forests of the Upper Midwest, which they had helped cut over after the Civil War. Appalachian logging peaked in 1909; that year 40 percent of the country's hardwood lumber came from the southern highlands.[30] Mountain logging was characteristic of a region-wide southern commodification of the woods during the same decades. Geographer Michael Williams has called the movement of northern logging companies south between 1880 and 1920 a "northern invasion," and he argues that it was "a classic case of an underdeveloped region where the first industries to be developed were extractive and exploitative."[31] Timber had long been an Appalachian commodity, but the scale of early-twentieth-century logging was entirely new.

Not even the most rugged or scenic locations were spared the lumberman's saw, though the intensity and timing of logging in various reaches of the mountains varied widely. In the Great Smoky Mountains, large timber corporations arrived around 1900, displacing many local operators, and cutting accelerated rapidly thereafter. The most famous of these operations was the Champion Fibre Company, a Cincinnati business that opened a pulp mill in Canton, North Carolina. Shortly after its construction in 1905, the Champion Mill "became the largest such operation in the world."[32] Exploitation of the even more rugged Black Mountains came a bit later. Much of the valuable lower elevation hardwood had been cut at least once by 1909, and after that northern corporations timbered the red spruce and Fraser fir of the range's upper reaches.[33] Throughout the Appalachian range, as early as 1902 a government report could lament that both native and absentee lumbermen had wastefully savaged the forests, with "no thought for the future."[34]

In the 1890s biologist and writer Margaret Morley visited the mountain, investigating the state of Grandfather's forests and recording her observations. Morley had come to western North Carolina from New England and was so enamored with the region that she remained for more than a decade. Unlike Dugger, Morley objected to scenic tourism out of a fear that it might alter or

FIGURE 6.3. Girdled trees, with Grandfather Mountain in the background. Photo in U.S. Department of Agriculture, *A Message from the President of the United States Transmitting a Report of the Secretary of Agriculture in Relation to the Forests, Rivers, and Mountains of the Southern Appalachian Region* (Washington D.C.: GPO, 1902).

harm Grandfather's landscape. She noted that the mountain had escaped systematic logging, but she was still troubled by the small-scale timbering in evidence as she hiked up its slopes. Characterizing Appalachian agricultural practices as primitive and unaesthetic, Morley described deadenings (plots where farmers had girdled trees to let in sunlight for farming) as "dreary openings in the forest" that were "made and abandoned one after another as the thin soil wears out, which on the poorer slopes happens in a year or two" (see fig. 6.3).[35] Worse still was the selective cutting of prized tree species on the mountain. Morley lamented that loggers had felled almost every mature wild cherry tree, leaving only young saplings in the coves. Despite her critiques, Morley seemed to be overwhelmed by nature as she climbed higher up the ridge; she wrote of a wonderland of wild fruits, nut trees, and blooming herbaceous plants at every turn.[36] Morley's 1913 book *The Carolina Mountains* was an admonitory tale, cautioning the reader that a lack of environmental protection stemmed from a failure to appreciate nature and suggesting that local uses of the mountain and its resources were unenlightened and wasteful.

As Grandfather was among the most rugged of all Blue Ridge peaks, timbering on the mountain came unevenly, over many years. When a Department of Agriculture report from 1902 cataloged the state of Grandfather's forests, no clear-cut timbering had yet taken place. The authors noted, however, the same selective cutting of certain valuable species, such as black cherry, chestnut, and red oak, that Morley had documented. The report described the eastern and southern slopes of the mountain as "lightly timbered," while the western and northern sides had "been somewhat culled, but [were] still heavily wooded."[37] Photographs included in the report are even more informative. The vista from the summit of the mountain reveals some small clearings, and another photo taken downslope displays a farmstead with rough fields laid out in the shadows of girdled trees.[38] For all this evidence of human labors, the report's authors described the upper reaches of the mountain as fundamentally wild. The spruce and fir forest near the peaks were "virtually primeval . . . an undisturbed forest equilibrium."[39]

Hidden from view from the peak, however, the woods several miles below the "primeval forest" were in the midst of a fundamental change. The founding of the resort town of Linville (along the river of the same name) at the southern end of Grandfather in 1891 marked a watershed in the history of the mountain. According to local lore, in 1888 Hugh MacRae, the wealthy son of Wilmington shipping magnate Donald MacRae and a graduate of the Massachusetts Institute of Technology, rode through the northwestern North Carolina mountains in search of mineral rights to purchase. MacRae sought deposits of iron, coal, or mica, preferably at low prices, seeking a North Carolina equivalent of the mining opportunities of southwestern Virginia or the Pocahontas coalfield. Other accounts describe multiple, more calculated, scouting ventures to determine a suitable mountain site for timbering as well as attracting tourists, backed by the MacRae family and other investors.[40] Struck by the recreational and tourist-attracting potential of the Grandfather region's rugged scenery, Donald MacRae and resort town developer Samuel T. Kelsey purchased approximately sixteen thousand acres from Walter W. Lenoir—a prominent planter, former slaveholder, and promoter of agricultural intensification in the mountains—that included the high reaches of Grandfather Mountain and several surrounding valleys. Hugh MacRae (who would manage the family investment) and Kelsey then set about planning a luxury resort town and large hotel that would cater to coastal and Piedmont residents looking to the cool air of the mountains for an escape from sweltering southern summers.[41]

Kelsey had a history of developing such vacation towns. His preceding project was Highlands, North Carolina, in the mountains south of Asheville, which quickly grew into a major southern vacation destination for the wealthy following its founding in 1875. Although he lacked Kelsey's practical business experience, Hugh MacRae was intensely interested in the connection between human nature and landscapes. Throughout his life he pursued ventures that combined multiple uses of the environment; his projects often joined timbering, resort development, real estate sales, and agriculture. MacRae also saw himself as something of a philanthropist, and, like many Progressives, he became interested in the idea of bettering people by placing them in healthy and productive environments.[42] Linville would reflect this belief in the restorative powers of nature, but his ideas found other expressions than just promotion of his mountain resort. By 1910 MacRae had invested in an eastern North Carolina agricultural community for European immigrants, convinced that such an agrarian village could foster cultural assimilation. He promoted new machinery and agronomic techniques, praised Henry Ford's paternal treatment of automobile factory workers, and awarded cash payments to couples on their marriage and the birth of a child. One 1916 promotional pamphlet captured his grand ambitions for his various land schemes in its title: *Vitalizing the Nation and Conserving Human Units through the Development of Agricultural Communities*. Although Linville lacked an explicit agricultural component, MacRae seemed to envision a similar sort of symbiosis between landscape and residents, in this case an environment of scenic recreation and a population of city-weary elites and middle-class professionals. If his agricultural community was supposed to strip away Italian or Greek identities and replace them with an idealized Anglo-Saxon American culture, Linville might also wash away the dust of Philadelphia's industry and substitute healthy and wholesome recreation.[43]

The village that grew at the foot of Grandfather was constructed in a uniform style, perhaps modeled on the large mountain "cottages" of Kelsey's Highlands project, Swiss mountain resorts, or influenced by the summer "camps" of the Adirondacks or the hay fever resorts of the White Mountains and Mackinac Island. Targeting discerning purchasers of second homes, modern houses were constructed to look rustic, finished in shingle siding cut from chestnut bark. They lined the Linville River bottom a few miles south of Dugger's hotel, many with views of the stream, Grandfather Mountain, or a new golf course that was part of the development. The crowning structure was an impressive hotel, the Eseeola Inn, designed in a similar architectural style. One visitor from Raleigh would later describe the village that emerged as "a gem of a summer resort, with beautiful hotels and dwellings, all in the Swiss style."[44]

Advertising was a crucial ingredient in the success of mountain vacation towns, and the resort management, the Linville Improvement Company (LIC), sent national publications descriptive pamphlets with such titles as "Western North Carolina, the Most Beautiful Mountain Region of the Continent" and "News from a Summer Land." They touted the clean air, stunning vistas, and serenity of the North Carolina highlands.[45] The LIC used nature to advertise the town as a recreational getaway but was a bit concerned about exactly what sort of nature visitors might encounter. MacRae declared that the path to success resided "in making Linville a place of beauty and a popular resort for health and pleasure for the best class of cultivated people possessed of means to aid in adorning and beautifying the valley."[46] Early promotional literature implied, however, that there was some beautifying work to do. Linville was apparently erected on a tract of land that had seen recent intensive logging. An advertisement of town lots for sale circulated by the LIC in 1891 featured only a rough-cut stump with the silhouette of Grandfather Mountain in the background.[47] The same year, Shepherd Dugger described the town site as "cleared and stumped so clean that it looked like a desert bordered with trees."[48] Despite these aesthetic shortcomings, the developers envisioned Linville as an opportunity to make a new nature from a clean slate, one ordered and organized to promote healthy recreation and civilized diversion.

As mentioned above, the town's best advertisement came through Dugger's novel. As part of the festivities and the promotional campaign accompanying the first season's Fourth of July celebration, the LIC held a contest for a work of fiction set in the Grandfather region, with a cash prize of $1,000. Although the winner ("In the Afterglow," by Isabella Maud Rittenhouse) went unpublished, Dugger's *The Balsam Groves of the Grandfather Mountain* was one of the entries.[49] A literary contest might seem a puzzling choice for drawing the attention of second-home buyers, but it was in keeping with popular portrayals of the southern mountains at the time. The works of writers like John Fox Jr. and Mary Noailles Murfree drew attention—and tourists—to particular portions of the mountains, even as they helped form a stereotypical Appalachian identity. Other communities had successfully leveraged works of fiction that also doubled as tour guides, such as Asheville's promotion of Frances Fisher Tiernan's *"The Land of the Sky," or, Adventures in Mountain By-Ways*, first published in 1875.[50] Linville's investors perhaps hoped that the contest winner would generate a best seller and thus a handsome return on their investment in prize money, and in some ways Dugger's novel did so, as it would go through four editions and remain in print for more than forty years.[51]

Dugger's novel followed many of the conventions of Murfree's and Fox's works. *The Balsam Groves of the Grandfather Mountain* presented the local inhabitants in much the same light as the natural aspects of the mountain: as curiosities for the adventurous tourist to observe. A description of the character Skipper John Potter is particularly revealing. Potter is a man with knowledge "so limited that the lack of it was by him unmissed." When he encounters a group of tourists sojourning on Grandfather, it is "the first time that [he] had fallen in with well-dressed, intelligent people."[52] Dugger portrayed locals herding cattle, hunting game, or simply attending to household necessities in small, isolated cabins where they "made love, soap and hominy all in the same room."[53] Throughout the novel Dugger uses the appearance of mountaineers pursuing daily activities as enjoyable, diverting entertainment, inserted to emphasize the wild, rustic nature of the region. Their exceptionalism serves the same literary purpose as descriptions of tall trees, rocky cliffs, and spectacular sunsets.

Dugger's fictional portrayal of Appalachian people to some extent repeats the stereotypes of scientists like Elisha Mitchell and Asa Gray, who had visited the mountain in the early and mid-nineteenth century, but it is also typical of a substantial body of writing on Appalachia at the end of the nineteenth and beginning of the twentieth centuries. Accounts such as Will Wallace Harney's "A Strange Land and Peculiar People" (1873), Murfree's *In the Tennessee Mountains* (1884), William Goodell Frost's "Our Contemporary Ancestors in the Southern Mountains" (1899), and Fox's turn-of-the-century novels all emphasize the isolation of Appalachian people and characterize mountaineers as primitive Anglo-Saxons forgotten by time and progress and in need of outside salvation.[54] Alluding to Murfree's popular descriptions of mountaineers, one early Linville ad described a land where "dwellers in the tiny clearings—types of Craddock's charming novels and many others yet unwritten—still move calmly through their secluded lives; and men in homespun clothes, who have never seen a railroad, still hunt the bear and deer through gorge and pass with ancient flint-locks."[55] While the characters in *The Balsam Groves of the Grandfather Mountain* are in many ways classically stereotypical, they also seem more than simply products of geographical isolation. Characters like Potter are also children of nature, direct reflections of the magnificence of the environment, and an allusion to the curative and fortifying power of direct contact with an untamed outdoors. These are people truly in touch with their environment.[56] It would be easy to take the characterizations of local Appalachian people as a distinctive phenomenon too far. Tourists to the White Mountains, Denver, and Arizona also enjoyed representation of caricatured rural Yankee, cowboy, and Hispanic cultures as part of their vacations.[57] "Different" people had the power to symbolize different—and potentially healthier—landscapes as much as they had the power to confirm a tourist's superiority.

Later expanded editions of Dugger's work devoted more space to pitching Linville as a vacation destination for sportsmen, promising that from "Linville's rustic houses" they could enjoy fishing for hatchery-raised trout protected "against the dynamiter and the clandestine angler."[58] Other promoters also emphasized these opportunities. The memoir of John Hartley, a local resident employed by the LIC in various capacities, noted the pleasant prospect for Linville sportsmen in the 1890s, stating that "the streams were full of brook trout and there was plenty of wild game."[59] (The resort drew on a tradition of local angling. Prior to Linville's founding a circuit of local streams and network of residents accustomed to housing wealthy anglers had developed along the Watauga, Elk, and Linville Rivers and was advertised in national sporting publications like *Forest and Stream*.)[60] Most Linville guests also partook of another outdoor pastime: golf. The resort built a course within its first five years, and the game became quite popular in the first decades of the twentieth century.[61]

The LIC continued to produce a number of pamphlets and brochures that touted the area's scenic vistas, with Grandfather serving as the dramatic centerpiece. In typical examples, promoters praised "endless original forests, unsurpassed for the beauty and variety of their trees" and, despite the presence of local logging operations, described a "particularly attractive mountain and forest land, owned and protected by the Linville Improvement Company," asserting that "nowhere have the beauties of nature been better preserved."[62] Even John Muir, the fiery and famed proponent of leaving America's wild land undeveloped, was captivated by the LIC's vision. In an 1898 tour of the Appalachian Mountains that included a visit to Grandfather, Muir found the view from the summit—"ridge beyond ridge, each with its typical tree-covering and color, all blended with the darker shades of the pines and the green of the deep valleys"—so beautiful that he burst into song.[63] Akin to the government forest surveyors of the same era, Muir imagined the mountains surrounding the growing resort to be essentially primeval.

Like Dugger and the operators of other mountain resorts, the LIC also pitched Linville's purportedly healthy climate and fresh air. It was, promotional literature declared, "an excellent summer residence for those afflicted with hay fever."[64] Another ad assured readers that, in addition to glorious scenery, the valleys around Grandfather were neither too hot nor too cold; instead, it was a landscape where "the invalid finds but few days uncomfortably cold for out-door exercise." As a consequence, "for many years past tired, nervous people, and those suffering from malarial poison and throat and lung troubles, have frequented this region and have, we believe, been invariably benefited." Seeking to seal the deal, the promotion promised that the resort was particularly suited to urbanites looking to escape city pressures in the southern mountains: "Literary people and brain-workers, especially, will find Linville suited to their wants. The altitude gives an invigorating atmosphere which acts as a tonic and goes far toward overcoming fatigue and making life profitable and pleasant."[65]

If scenery and climate were indispensable elements in building an attractive nature resort, reliable transportation was just as crucial to its financial success.

Much as industrial centers like Roanoke relied on an expanding railroad network to fuel their growth, vacation towns also depended on the rails and roads for their increasing popularity. Since the resort site initially had no convenient connections to the lowlands, in 1889 the LIC began construction of a toll road, named the Yonahlossee (perhaps after a Cherokee word), winding along the eastern flank of Grandfather Mountain to connect the prospective resort to the town of Blowing Rock. A crew of a hundred workers blasted and dug the road through rocky, treacherous terrain, relying on human muscle and mule power to build a section roughly twenty miles long.[66] Although it was constructed to satisfy the needs of a single company, observers consistently portrayed the Yonahlossee Road as a herald of coming regional prosperity and praised its engineering and scenic views in print. One traveler noted the road's "commanding constant views of peerless scenery," while a local engaged in the tourist trade declared it "a perfect road in all respects, built on practically a level grade and well made and maintained."[67] All parties were convinced that the Yonahlossee Road was a key ingredient in the region's future success.

Good local roads may have been the final link in visitors' journeys to Linville, but they rested on a foundation of an expanding southern rail network. Other southern mountain tourist sites had boomed with the arrival of the railroad, and Linville's boosters also looked forward to the day—presumably soon—when their town would become an important rail stop.[68] By the 1880s a rail line from Tennessee came within ten miles of the proposed village, and Linville promotions promised that an extension to the resort was forthcoming. Tourists coming from the north or west could, by the time the resort opened, take the Southern Railway and then the East Tennessee and Western North Carolina Railroad to the station at Montezuma, within two miles of Linville, or if coming from the south and east they could take the Carolina and Northwestern Railway to Edgemont, and from there take a new stage road up the mountain to their final destination.[69]

While the LIC counted on natural beauty, a healthy climate, and improving transportation networks to attract guests and home buyers to the resort, management was also intent on extracting profit from the company's holdings though other ventures. When the East Tennessee and Western North Carolina Railroad finally extended its line to Linville around 1900, in addition to carrying tourists the trains provided lumber companies easier access to the mountain's timber resources.[70] As a result, the first two decades of the twentieth century saw the exponential growth of intensive logging on Grandfather. The LIC operated its own sawmill on the edge of town, where it harvested the "original growth of spruce, balsam and hemlock," and the company sold timber rights to bigger operations on the other slopes of the mountain.[71]

FIGURE 6.4. "Yonahlossee Road on the Southern Slope of
Grandfather Mountain," circa 1902. Photo in U.S. Department
of Agriculture, *A Message from the President of the United States
Transmitting a Report of the Secretary of Agriculture in Relation to the
Forests, Rivers, and Mountains of the Southern Appalachian Region*
(Washington D.C.: GPO, 1902).

The largest of these outside organizations, Whiting Lumber Incorporated,
opened the Boone Fork Lumber Company mill in the growing town of Shull's
Mill at the northern end of Grandfather in 1915, and it began building narrow-
gauge railroad lines up the heavily timbered hollows and ridges. By 1917 the
company had three hundred men employed cutting spruce, ash, oak, poplar,

and chestnut in the remote coves. In 1918 alone the Boone Fork Lumber Company clear-cut 1,436 acres on the mountain and milled more than 1.6 million board feet of lumber.[72] The LIC also focused on extracting a profit from the more rugged, upper reaches of the mountain. The company issued the first contracts for cutting the upper-elevation red spruce, the "primeval" forest that government officials had noted during their 1902 visit, in 1918. A year later the majority of the mountain had already been cut over. In an account for stockholders, LIC officials calculated that there was only $400,000 worth of standing timber remaining on their holdings that might still interest logging companies.[73]

Although logging and the sawmills provided employment for local residents, their environmental impact was not popular with everyone. One claimed that the denuded sections of the mountain resembled a "moonscape of stumps and mud."[74] Significant fires, fed by the piles of limbs, bark, and treetops left by logging operations, did further damage to the scalped clear-cuts (the most devastating took place in 1926 and 1940), and soil-eroding floods regularly occurred without the forest canopy and root networks to slow runoff from rainstorms.[75] By the early 1930s, loggers were entering the last untouched groves of timber on the mountain, spurred by a high demand for red spruce in the aviation industry. Timber companies built sixteen miles of plank road to access the tops of the tallest ridges, and by 1936 they had completed clear-cutting the accessible portions of Grandfather.[76]

In addition to timber and scenery, other forest resources proved economically important during the resort's first decades. A successful nursery operated by Harlan Kelsey brought Linville to the attention of the nation's gardeners, marketing many of the same rare highland species that had attracted botanical collectors to Appalachia a century earlier. Kelsey, the son of town developer Samuel Kelsey, opened Kelsey Highland Nursery in 1889, before the resort village was much more than an idea. A decade later he moved his office to Boston, but the Linville site remained the nursery company's propagation center. Kelsey also became a noted landscape architect and lobbied for a variety of campaigns, including the preservation of colonial architecture and maintaining New England's scenic vistas by cleaning up billboard clutter along roadsides.[77] Kelsey spent much of each year in the town his father helped construct, and he came to love the flora of the southern highlands. Combining his training in landscape architecture with this botanical interest, Kelsey's nursery specialized in rare specimens, from showy azaleas and Carolina Hemlock to *Stuartias* and dwarf Allegheny sand myrtle, advertising the High Country's plants in regional and national publications.[78]

Concerns about timbering led Kelsey and others who valued standing forests and scenery more than lumber profits to advocate for some sort of protection for Grandfather Mountain and the surrounding ridges. These efforts, with Kelsey as the driving force, came close to establishing the Linville area as a unit of the National Park Service (NPS). He had developed an intense interest in southern Appalachian preservation by the 1920s, and his fondness for Grandfather led to promotion of the Linville region when NPS officials were considering the best locations for an eastern national park. (The mountain was also under contemporaneous consideration as a waypoint on the nascent Appalachian Trail.) By this point in time, Kelsey was in a position to wield some influence: in 1897 he had joined the Appalachian Mountain Club, an influential private organization focused on preserving New England's mountains from logging, eventually served as its president, and communicated regularly with important NPS officials such as Stephen T. Mather and Horace Albright. In 1924 Kelsey was selected by the NPS as one head of a Southern Appalachian National Park Commission, tasked with locating the best site for an eastern national park. As part of a fact-finding tour, the commission members visited several sites in North Carolina and Virginia, including Grandfather, the location Kelsey initially supported.[79]

Despite Kelsey's advocacy, the High Country would ultimately lose out to two other southern Appalachian sites: a stretch of the Blue Ridge Mountains east of Virginia's Shenandoah Valley, and the Great Smoky Mountains in southwestern North Carolina and eastern Tennessee. National park officials found the Virginia tract's proximity to major eastern cities and a plethora of nearby historic sites appealing. The Smokies benefited from a confluence of factors: the growth of Knoxville and Asheville (where the most vocal southern national park booster organization, the Appalachian National Park Association, had been based since its founding in 1899); a larger swath of cheap acreage than was available in northwestern North Carolina; local activism stimulated by concern about the voracious timber appetite of the Champion Fibre Company; and promotion by figures like Horace Kephart, a travel writer whose book *Our Southern Highlanders* (1913) brought widespread attention to the Smokies as a unique natural and cultural treasure. The Great Smoky Mountains National Park opened in June 1934, and Shenandoah National Park opened the following year.[80]

Although Kelsey was unsuccessful in establishing the Linville area as a national park in the 1920s, he would not surrender the attempt easily. He would try again in the 1940s to arrange sale of Grandfather Mountain and the adjacent Linville Gorge to the National Park Service, fearing for its future and still convinced that it would make a fine park. After nearly a decade of wrangling—first with Hugh MacRae, then MacRae's son-in-law Julian Morton, and finally a grandson, Hugh Morton—and a pledge of half the purchase price from philanthropist John D. Rockefeller Jr., neither Congress nor North Carolina were willing to spend public funds on the mountain (though Rockefeller would purchase the gorge and grant it to the public). Negotiations broke off without an agreement as to the price or the ultimate purpose of the mountain. Kelsey, as a pseudo-representative of the NPS, envisioned a national park unit focused on environmental preservation and outdoor recreation, with Grandfather Mountain as the centerpiece, while the MacRaes and Mortons wished to continue their tourism, real estate, and timbering ventures unless the federal government was willing to make them an offer they could not refuse. Despite Kelsey's prodding, NPS officials refused to invest in the Linville area, and the negotiations collapsed once and for all.[81]

Despite the failure to secure national park status, by the mid-1930s Linville Resort was well established and slowly growing, despite the economic tumult of the Great Depression and a moderate debt burden.[82] The LIC and the MacRae family continued to place an emphasis on the health benefits of vacationing in the highlands, especially the relaxing effects of time spent outdoors in the mountains, and improved gravel roads coupled with the rising popularity of automobiles led tourists to drive to Grandfather in ever-increasing numbers. Most contemporary visitor accounts failed to mention the clear-cut and burned-over bulk of the mountain but elaborately described the beautiful environs immediately surrounding the resort. In 1935 the national travel magazine *Touring* represented Linville as an invigorating outdoor recreation wonderland "created for zestful living." A typical stay in the bark-sided hotel included access to a skeet range, horse trails, a golf course, trout fishing, tennis courts, and the wild forests of Grandfather (at least in the few spots that remained uncut). An ad in the same issue sold the combination of natural scenic beauty and tasteful development, marketing the town as "truly rustic yet thoroughly modern."[83]

In the early 1930s, the LIC built a toll road part of the way to the top of Linville Peak on the southern end of the mountain. This road-building on Grandfather mirrored extensive highway construction in America's national parks and forests during the Depression years under the auspices of various public make-work projects.[84] At a cliff-side overlook, the company constructed an observation deck and a gift shop and began charging a toll of fifty cents per vehicle.[85] This "scenic attraction" was immediately popular, reinforcing the notion that the visual consumption of nature was an avenue to economic gain. And this success was intimately tied to swelling automobile tourism. As historian Allen Batteau has claimed, it was the car "that made possible the use of Nature as a commodity for recreation."[86]

Increased automobile ownership and travel, coupled with this new ease of access, brought many more tourists to the upper elevations of the mountain,

FIGURE 6.5. "Auto Road along Grandfather Mt. (Ele. 5964 ft.) Linville, NC." Photo by Howard Pelton, 1912. Courtesy of the Library of Congress.

and newspaper articles published in the state's Piedmont cities touting the natural beauty of Grandfather provided free advertising. A visitor from Charlotte exemplified the boosterism of many descriptions from the 1930s. In 1939 he declared that the mountain had an "Edenic simplicity and sublimity" and was "one of the select specimens of creative carving in the Almighty's museum of sculpture."[87] Another traveler, from Greensboro, remarked on the ease with which visitors could access the summit via the new gravel road and a footpath and was amazed at meeting foreign tourists on the slopes of the mountain.[88] By 1937 the toll road was showing a promising return, and LIC president Nelson MacRae advocated extending the road to the summit of the mountain and making it the centerpiece of Linville's ad campaign.[89]

Neither the newspaper articles nor the 1930s Linville advertisements mentioned the familiar trope of charmingly ignorant locals. Mirroring the promotion of national and state park systems, the focus of the ads had shifted to nature and outdoor recreation alone. The attraction employed local residents, but they had become ticket sellers, fire wardens, or gift shop clerks, useful for their labor rather than their culture.[90] The success of previous scenic advertising and the money invested in the toll road and outdoor recreation activities fueled an ever-increasing focus on nature tourism, with less and less emphasis on cultural distinctiveness. The late 1930s was also a time of growing public interest in seeing the nation's natural wonders and considering the importance of "wilderness."[91] When cultural performances did appear, they often had only tangential ties to older ideas about locals. Grandfather's most popular "folk" exhibition would become annual Scottish games, purporting to celebrate the region's Highland clan traditions.[92] The LIC was selling nature and seemed to have little interest in propagating negative or backward perceptions of the region. There was no room for hillbillies in this new recreational landscape.

&

Even the High Country, one of the most remote regions of Appalachia, was caught up in the economic development sweeping the mountains following the Civil War. As in many parts of Appalachia, the timber industry proved the most transformative activity during this time, simultaneously bringing jobs, money, and environmental degradation (just as coal transformed other stretches of the mountains). Although late-nineteenth-century timbering would bring a new degree of social and environmental change, human activity had already altered or modified the Grandfather region to a significant degree. Recent Appalachian historiography has emphasized the extent to which the southern highlands were connected to outside regions through trade, agriculture, tourism, and industry during the nineteenth century. Appalachia had never been truly isolated, and the increased construction of highways and railroads after the Civil War made travel throughout the region faster and easier.

Linville's story also illustrates the changing economic values that came to characterize large stretches of the southern mountains by the early twentieth century. Individuals like Shepherd Dugger, Hugh MacRae, and Harlan Kelsey saw value (economic and otherwise) in the region's scenic beauty and imagined salubrity. Logging and mineral extraction, while remaining important throughout the mountains, lost ground to tourism in certain locales, though it was a patchy and gradual process. Glowing descriptions of the beauty of certain peaks, the healthiness of particular valley towns, and the comfort of specific accommodations lured more and more people to Appalachia, intent on vacationing in a scenic and restorative place.

Linville's model of private development of scenic sites would be a common one in twentieth-century Appalachia, as a wide range of developers and investors cashed in on vacationers' desire to consume scenery, enjoy outdoor recreation, and bask in healthy environments. Sites like Little Switzerland, bordering the Blue Ridge Parkway, followed Linville's model of advertising and a distinctive architectural theme (in its case pseudo-Swiss). Chimney Rock lured tourists with its spectacular stone formation and commanding views, and the Vanderbilts' mountain estate, Biltmore, was transformed into an attraction that draws tens of thousands of visitors each year, in part because of its extensive, park-like grounds and forest.[93] And Grandfather Mountain's joining the state park system of North Carolina and its close relationship with the Blue Ridge Parkway mirrored other fusions of private and public interest in mountain environments. Outdoor sports companies today sell gear and provide river rafting and rock climbing experiences that rely on state and federal lands. Hotels and cabin rental businesses depend on the attraction of nearby national parks and national forests. Chimney Rock, like Grandfather, has sold much of its interest in land to the state of North Carolina, while maintaining a concession license on-site.[94] The most famous regional public-private hybrid attraction is surely the Appalachian Trail. Begun in the 1920s as a planned footpath along the crests of the mountains from Georgia to Maine, the trail was originally plotted by private interests, though it eventually fell under the auspices of the National Park Service. The trail makes its way over a combination of public and private lands, with much of its maintenance provided by volunteer groups.[95]

Today the fusion of private and public continues at Grandfather. The federal Blue Ridge Parkway cuts across the mountain's eastern flank, handling millions of cars each year. Below the parkway, the Pisgah National Forest attracts hikers, campers, and hunters. Since 2008, the North Carolina Division of Parks and Recreation manages the bulk of the higher slopes of the mountain, protecting its forests and rocky outcrops from development. And a private attraction remains on the end of the mountain adjacent to Linville (still a quiet but wealthy community of second homes inhabited by golfers and fly fishers). An outgrowth of the LIC's original toll road and scenic overlook, the modern tourist site maintains a "mile-high swinging bridge," a small zoo stocked with native animals, and a nature museum. A nonprofit corporation administers the attraction and promises that "all proceeds from sales of tickets and souvenirs go toward caring for and presenting Grandfather Mountain in a manner that inspires good stewardship in others."[96]

In the High Country, and in much of Appalachia outside of the coalfields, outdoor recreation tourism has supplanted logging, agriculture, and other pursuits as the most important economic activity. Small farms, logging camps, and narrow-gauge railroad beds have been replaced by second homes, scenic parkways, and zip lines. Tourists enjoy both public and private landscapes, still convinced of the value of scenic beauty and the idea of a "healthy" vacation. Today those tourists are more likely to enjoy hiking, rock climbing, kayaking, or cycling than a century ago, but they still retain a notion of the restorative power of beautiful views and fresh air. Together public lands, private developers, and scenery-seeking tourists have transformed large swaths of Appalachia into outdoor playgrounds. These transformations that people began at places like Linville have steadily divided the southern mountains into two Appalachias, one a rural-industrial region of strip-mining, embattled unions, opioid abuse, and poor hollows, and another of ridgetop vacation homes and pristine nature (or at least advertisements of such). The consumption of scenery as surely as coal contributed to this division.

CHAPTER 7

Tobacco

*Making Ground for
an International Crop*

THE 1998 MASTER SETTLEMENT AGREEMENT (MSA) between the largest U.S. tobacco companies and the majority of state governments promised the end of a centuries-old pattern of Appalachian farming. The deal quashed a range of public health lawsuits against the companies in exchange for tremendous annual payouts to the states. It also led in 2004 to the Tobacco Transition Payment Program (TTPP), which ended existing federal regulations of raw tobacco production and sales, laws that had kept small mountain farmers producing tobacco for years. To be sure, Appalachian tobacco fields did not disappear overnight. A drive through rural hollows from eastern Kentucky to western North Carolina today can still furnish views of small fields of burley tobacco, farm families cutting plants, and barns full of drying leaf, depending on the month. Slowly but surely, however, the MSA led to smaller crops, fewer tobacco farmers, and the dissolution of an Appalachian culture more than two centuries in the making.[1]

When lawyers hammered out the agreement, tobacco had long been the cash crop that connected mountain farmers to distant markets. Even an acre or less of tobacco meant a crop intended for more than just home use. And this cultivation was not an unchanging continuation of the same form of agronomy year after year. Appalachian tobacco farmers kept abreast of contemporary practices across the nation, experimented with seed varieties, explored promising cultivation and fertilization techniques, and adopted new technology when they believed these things would increase production and profits. Farmers embraced new varieties of tobacco when they became popular, participating in booms in bright leaf in the late nineteenth century and white burley in the early twentieth century. Despite stereotypes that mountain farms had long been isolated and insular, tobacco growers proved enterprising and innovative individuals who looked to the future as much as to tradition.

A history of the crop reveals the tensions between physical landscapes and environmental ideals in mountain farming. On the one hand, tobacco growing depended on a farmer's intimate knowledge of specific environments. Growers and boosters linked high-quality tobacco "types" like bright leaf and white burley to specific climates and soils, finding examples of these ecotypes in various corners of the mountains. At the same time, farmers and tobacco manufacturers strove to define each type as a commodity, something they could sell and trade by name and grade across vast distances, sometimes on description alone. The contradictions of these two concepts—an environment-specific crop but one that was traded across vast distances as a fungible good—remind us of the ironies inherent to some extent in ideas about all Appalachian natural resources turned commodities.

Mountain tobacco growers became enmeshed in more than private networks of exchange. Before the Civil War, state agents inspected tobacco at warehouses, certifying its quality. Bulk leaf and manufactured tobacco products traveled on state-regulated railways. By the late nineteenth century, land grant universities and state government employees conducted and published a range of tobacco studies and tested fertilizers for purity. And then, during the New Deal, the federal government intervened in Appalachian growers' lives in new ways, monitoring soil conservation, studying rural communities, and most directly by limiting tobacco acreage and establishing price supports. This federal tobacco program bound mountain fields to tobacco crops in Florida, the coastal plain of the Carolinas, and the Maryland Chesapeake, as well as international markets in leaf, more tightly than ever before, even as it ensured tobacco's future in Appalachia.

The TTPP finally eliminated the old regulations and supports that originated in the New Deal and changed the relationships between mountain farm environments and ideas about their productivity. Appalachian farmers were once again thrown into direct competition with national and even global growers, and the math had changed. Companies sought efficiency through fewer farmers and fewer contracts, pushing economies of scale over diversity in production. Growers in the flat, sandy fields of eastern North Carolina and South Georgia increased their bright leaf acreage at the expense of smaller farmers in the western Piedmont, and big burley operations in central and western Kentucky grew at a cost to small farms in the mountains. Today, looming over all of these rural communities, is the fear that some distant land, in, say, Brazil or China, will be the new tobacco frontier, ready to supply cigarette makers with the newest varieties cheaper than can any U.S. farm. Tobacco has fully entered the global world of agribusiness, among the last U.S. crops to undergo this revolution.

ളാ

From the earliest days of colonial settlement, most Appalachian white and black residents engaged in some sort of farming, and many of those farmers produced at least a small surplus for market. Farmers could sell or barter excess corn or wheat (and especially the whiskey distilled from those grains), fleshing out larders or furnishing small amounts of cash needed for obligations like taxes.[2] Livestock was the most important form of agricultural surplus in early Appalachia. By the late eighteenth century, southern mountain herdsmen supplied meat to the lowlands. Appalachian environments were particularly suited to raising stock. Thanks to cooler temperatures, abundant rainfall, and strong soils, grazing land was better than in the Deep South, and the climatic limitations circumscribed agricultural competition from cotton.[3]

As communities developed west of the mountains, droving proliferated. Mountain livestock walked west as well as east, and the mountains also came to serve as a highway for trans-Allegheny stock headed to coastal towns and plantations. Asheville became the great hub for this trade, but other communities heavily relied on livestock markets too. For example, in eastern Tennessee, Cades Cove lay at an intersection of cattle trails from the west and the south. Drovers stopped over in the valley to feed their herds on its grasses, and cove residents raised surplus cattle on the surrounding mountain balds to sell to these herdsmen, who then drove them on to Knoxville. From there, routes led northeast through the Great Valley to eastern urban centers.[4] In addition to cattle, mountain farmers also sent substantial numbers of hogs, horses, mules, and sheep to market prior to the Civil War.[5]

This commerce makes clear that Appalachian people sought market outlets for some of their produce from the early days of settlement. A murkier question is exactly what sort of farmers they sought to be. Was their preference for a basic stake, independence, as the image of the Revolutionary-era Overmountain Men made popular, or was it to replicate the agricultural networks of the lowland East? One answer, of course, is that it depended on the individual. Some people sought solitude and self-reliance, while others pursued community and connection. No doubt for every "long hunter" who supplemented hunting and fishing with a small plot of corn for meal and a rangy hog or two, there was a family searching for improved stock and a place to sell surplus cattle and grain.

Debates over the prevalence of subsistence versus market agriculture obscure as much as they reveal. The two need not be mutually exclusive. Indeed, much of the rhetoric of the U.S. agricultural press of the first half of the nineteenth century dwelled on the desirability of farm families providing for their own needs while also producing specialized crops for market. These activities, rather than being at odds, were two sides of the same coin. Solvency came only from a balanced farm economy.[6] Most Appalachian farmers lacked access to these journals or, likely, had no desire to read them and apply their complicated precepts, and many small farmers were no doubt semiliterate or illiterate. But consistent antebellum lobbying for internal improvements throughout the region echoed similar pleas across the rest of the rural nation and seem to strongly indicate a desire for a more connected agricultural economy.[7] Farmers may not have read much reform literature, but their actions often mirrored its aims.

Although less important than livestock in the early years, another crop bound for distant markets characterized much mountain farming: tobacco. As the colonial southern backcountry expanded into the Blue Ridge of Virginia and North Carolina, tobacco planters carried their cash crop with them, with the result that the eastern edge of the mountains was sprinkled with farm communities similar to those of the flatter lands to the east. Mountain settlers who came from farm communities in the Tidewater or Piedmont were accustomed to growing the crop, and they also brought it with them to their new homes deeper in the mountains, where they often discovered it grew well. For example, pioneering families in Tennessee's Watauga settlement, home of the Overmountain Men, raised tobacco from the start.[8] And in the communities of central Kentucky, transplanted Virginians brought seed and cultivation practices from the Piedmont South into the new environment of the Bluegrass region and the western edges of the mountains.[9]

Some of this leaf certainly was for home consumption, but even a small patch of tobacco produced more than a family could use, encouraging engagement with local or distant markets. Some mountain farm communities incorporated tobacco production as an important part of their agricultural practices. By 1860 30 percent of Ashe County, North Carolina, farmers grew tobacco, for example.[10] More broadly, sociologist Wilma Dunaway calculates that mountain farmers raised more than thirty million pounds of tobacco in 1860, more than twenty-four million pounds in modern-day Virginia alone (though it is worth noting that her definition of Appalachia has been criticized as too extensive). Even farms in more remote mountain counties that lacked good river or turnpike access each averaged almost 240 pounds of tobacco for export that year.[11] Although most Appalachian leaf left the mountains as bulk produce, the region was also home to a growing tobacco manufacturing industry focused on adding value to the raw agricultural product, especially in the small towns of the Virginia and North Carolina Blue Ridge.[12] Engaged in raising one of the globe's most important staple commodities, Appalachian tobacco farmers could hardly avoid interaction with distant markets.

Appalachia's tobacco landscape would change with developments that first occurred beyond its margins in the Piedmont during the mid-nineteenth century. There, in a few counties along the border of Virginia and North Carolina, farmers experimented with raising and curing a form of tobacco that came to be known as bright leaf. Bright leaf's apocryphal history traces its beginning to a slave named Stephen on a farm in Caswell County, North Carolina. Stephen supposedly fell asleep while tending a barn of curing tobacco one night in 1839, and upon awakening and realizing that the curing fire had all but died, he added copious hot coals in an effort to compensate for his mistake. The resulting intense charcoal fire cured the leaf to a bright yellow, and it tasted mild and sweet. The tobacco sold exceptionally well, and Stephen's master, Abisha Slade, set about codifying a particular set of cultivation and curing techniques to reproduce the "mistake," becoming one of many boosters spreading the type through the counties of North Carolina's and Virginia's western Piedmont during the antebellum era.[13]

The Civil War paused bright leaf's spread, but in the years after 1865 the type continued to sell well, and boosters sought its geographical expansion. Eastern Piedmont growers raised larger and larger crops of bright leaf, and mountain farmers also grew interested in its potential. In the Blue Ridge, where farms had long produced fire-cured leaf, cultivators began to focus their efforts on bright leaf, with the result that pockets of counties like Stokes in North Carolina and Franklin, Henry, and Patrick in Virginia became important production centers. And boosters looked to the valleys of the higher mountains to the west as well, interested in exploring their suitability for growing bright leaf, especially as a spreading network of mountain railroads made shipping large tobacco crops more practical.[14]

Bright leaf's spread into the mountains not only replicated Piedmont tobacco culture, in some cases it was also driven by the same individuals. Some of the most prominent Appalachian boosters had experience with the crop in the older districts. J. D. Wilder, for example, had been involved in the Danville, Virginia, bright leaf market, and he came to Asheville to direct its first warehouse tobacco sale in 1879. The Reynolds family—hailing from the Blue Ridge of Virginia's Patrick County and famous for their role in making Winston-Salem a tobacco center—also established a tobacco manufacturing facility that converted bright leaf to plugs in Bristol, on the Virginia and Tennessee line, in the 1870s.[15] And one of the largest early bright leaf farmers in the countryside surrounding Asheville was W. T. Dickinson, who had been a well-to-do planter in Pittsylvania County, Virginia before moving to the mountains.[16]

The most influential transplant was Samuel Shelton. Before the Civil War Shelton helped pioneer bright leaf curing in Henry County, Virginia, from his plantation in Irisburg at the base of the Blue Ridge, and he published his studies of curing in influential agricultural journals, such as the *Southern Planter*.[17]

After the war, believing that Henry County's best days were behind it, Shelton moved to a farm outside of Asheville. In 1868 he grew his first mountain crop and set about trying to convince other tobacco farmers that Appalachia was bright leaf's new frontier thanks to the region's climate, soils, and abundant timber for curing fuel.[18] Shelton would go on to establish a small tobacco factory and a new curing system that he advertised to fellow growers. His voice became even more influential when he began writing a regular agricultural column in the *Asheville Citizen*, western North Carolina's most widely circulated newspaper.[19]

Proponents of tobacco's expansion into the mountains, like Shelton, offered convoluted advice about the crop's environmental requirements. On the one hand boosters noted tobacco's phenological flexibility, claiming that "no plant is so easily modified by climate, soil, and different methods of cultivation, as tobacco. Climate imparts flavor; soil determines texture."[20] On the other hand, they stressed that in order to grow the best examples of this malleable plant, soils and conditions had to meet demanding requirements, ones that many farmers might find counterintuitive. Unlike most crops, bright leaf demanded well-drained soils that were poor in nitrogen and relatively low in organic matter. Rich lands, or those that had been well manured or worked in careful crop rotation, produced heavy-bodied and harsh-tasting bright leaf. In essence farmers had to "starve" the plants in order to produce the light-bodied and yellow-colored leaf that consumers fancied.[21] The result, if farmers carefully followed the proper cultivation and curing techniques, was a tobacco to rival the fanciest "Turkish" grades.[22]

These odd agronomic requirements delighted boosters and mountain farmers alike. Although many mountain valleys contained rich bottom land, regional farms often also included sloped lands that were less productive for most crops. Since bright leaf grew well to 2,500 feet in elevation, growers could plant tobacco on hill land that might not have been useful for another crop.[23] J. B. Killebrew, a national tobacco expert, wrote that it was difficult not to marvel at the irony of a premium quality crop that grew best on the "no-soil, of these worn fields." This led to "[t]he strange spectacle . . . exhibited of lands which will not grow a bushel of wheat making net returns of from $300 to $500 per acre" when farmers planted them in bright leaf.[24] Experts touted particularly appropriate soils in various reaches of North Carolina and East Tennessee, stretches of decomposed sandstones or granites and old fields where "the conditions of sterility prevailed," turning a former agricultural liability into a lucrative asset.[25]

A speculative fever of sorts accompanied the spread of bright leaf west. Killebrew asserted that "the discovery of gold would not have conferred upon the

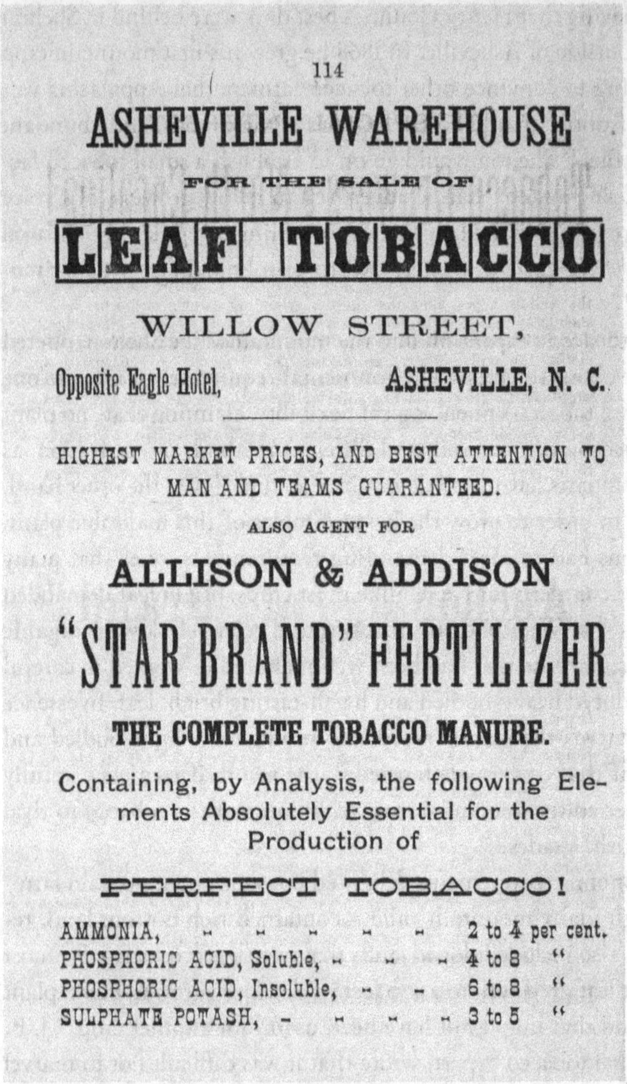

ASHEVILLE WAREHOUSE

FOR THE SALE OF

LEAF TOBACCO

WILLOW STREET,

Opposite Eagle Hotel,   ASHEVILLE, N. C.

HIGHEST MARKET PRICES, AND BEST ATTENTION TO
MAN AND TEAMS GUARANTEED.

ALSO AGENT FOR

ALLISON & ADDISON

"STAR BRAND" FERTILIZER

THE COMPLETE TOBACCO MANURE.

Containing, by Analysis, the following Elements Absolutely Essential for the
Production of

PERFECT TOBACCO!

AMMONIA,	2 to 4 per cent.
PHOSPHORIC ACID, Soluble,	4 to 8 "
PHOSPHORIC ACID, Insoluble,	6 to 8 "
SULPHATE POTASH,	3 to 5 "

FIGURE 7.1. Asheville became a center for bright leaf tobacco sales
and manufacturing in the late 1870s, and competing companies
like the Asheville Warehouse advertised for farmers' business.
Illustration from J. D. Cameron, *A Sketch of the Tobacco Interests of
North Carolina* (Oxford, N.C.: W. A. Davis, 1881).

poor regions of North Carolina and Virginia the present wealth or the hope of permanent industry afforded by yellow tobacco." He went on to compare the nascent boom with other contemporary southern industries rapidly converting the environment into commodities, equating bright leaf with the ironworks of the Alabama foothills and the growing cotton commerce of the Mississippi Valley.[26] In rural, mountainous Madison County in the late 1870s, a hundred-weight of bright tobacco could bring a farmer twenty dollars, in Wilkes some lots sold for forty, and in Buncombe "small lots, exceedingly fine," might sell for as much as $2.50 per pound. These were as lucrative prices as could be found in Caswell County, where bright leaf originated. In Asheville and other regional towns this growth meant new urban warehouses and factories as well as more rural tobacco fields. Just as the flow of coal through Roanoke connected rural mines and urban consumers through the commerce of the railroads, cured tobacco that passed through warehouses and plug factories created a nexus of farmers, town businessmen, and distant consumers.[27] The railroads were important here too, as western North Carolina's tobacco manufacturing was also linked to the advancing tracks of the Western North Carolina Railroad as well as the Spartanburg and Asheville line.[28]

Expressions of bright leaf's market-oriented nature could be found on rural farms as well as in urban warehouses and factories. Most plots of tobacco were small, an acre or two that might be tended by a farmer and his family to furnish ready cash for a farm that otherwise focused first and foremost on subsistence, but other operations were substantial even as early as 1880. Some exceeded twenty acres of tobacco land, demanding the labor of hired hands or tenants to make the crop. In 1880, for example, A. M. Alexander, of the French Broad River valley just north of Asheville, grew thirty acres of bright leaf; Weaversville's W. T. Dickinson raised forty-four acres; and J. M. Smith of Madison County cultivated fifty acres—what would have been considered a large tobacco crop in any part of the United States at the time.[29] The total harvest in Smith's county, a rugged and rural locale situated north of Asheville, that year equaled approximately one million pounds of cured leaf, worth an impressive $225,000 on the market.[30] This was hardly a crop for home use.

Mountain farmers also demonstrated their engagement with national tobacco trends through an avid interest in new techniques and technologies. Bright leaf boosters constantly released new seed strains, advocated particular plowing or cultivation practices, and furnished formulas for curing tobacco to a light yellow color. As the practices emerged, western North Carolina growers quickly embraced modern barn construction, installed sheet metal heating flues to replace open curing fires, individually harvested tobacco leaves as they ripened rather than cutting the entire stalk, and purchased blended chemical fertilizers to hasten ripening.[31] An 1880 census report noted that mountain growers practiced curing tobacco with metal flues at some of the highest rates in North Carolina, suggesting broad adoption of the technology.[32] Collectively the practice of rapid technological investment and adoption suggested a high degree of grower interest in and awareness of modern agronomic practices.

Despite—and in some case because of—their adoption of current cultivation techniques, mountain bright leaf farmers faced environmental challenges. Much Appalachian farmland was simply too rich to make good yellow tobacco, and the land that was suitably poor was often sloping and fragile. As a consequence, erosion was a constant concern. As was the case in the Piedmont, summer thunderstorms threatened to wash away the soil needed for next year's crop, carving hillside fields into mazes of gullies. Bright leaf's curing requirements exacerbated this problem, as farmers timbered more and more of their forested acres to feed the flues of their tobacco barns.[33] The most intensive studies of soil erosion accompanying bright leaf cultivation came from the western edge of the Piedmont. There one twentieth-century federal study observed that in two North Carolina counties "28 percent of the land had lost more than 75 percent of its topsoil. An additional 54 percent of the land had lost an average of at least 50 percent of its topsoil. Thus, less than 20 percent of the land had as much as three-fourths of its original topsoil left."[34] Research suggested that a similar situation existed in the portions of southern Appalachia, where yellow tobacco cultivation was concentrated.[35] Piedmont social structures as well as environmental challenges accompanied bright leaf as farmers expanded its culture westward. Tenants cultivated a substantial portion of the tobacco crop, despite the stereotype of independent mountain landowners, and regional tobacco manufacturers relied disproportionately on African American labor, as did their peers in the lowlands.[36]

Accompanying environmental and labor challenges, a convergence of economic factors weakened the bright tobacco fervor in the mountains by the end of the century. For one, competition among yellow-tobacco-producing regions grew ever fiercer. The old district of the western Piedmont remained important, but bright leaf had spread to other areas during the same years that it made inroads in the mountains. Phosphate fertilizers opened the sandy lands of the North Carolina Coastal Plain to high-quality tobacco cultivation. And, much as had been the case in the Asheville district, boosters from the old tobacco belt spread the crop to the Pee Dee region of northeastern South Carolina as well as fields along the Georgia and Florida line. Manufacturers bought more and more bright tobacco, especially as the popularity of cigarettes expanded, but every year farmers hauled a greater poundage of bright leaf to

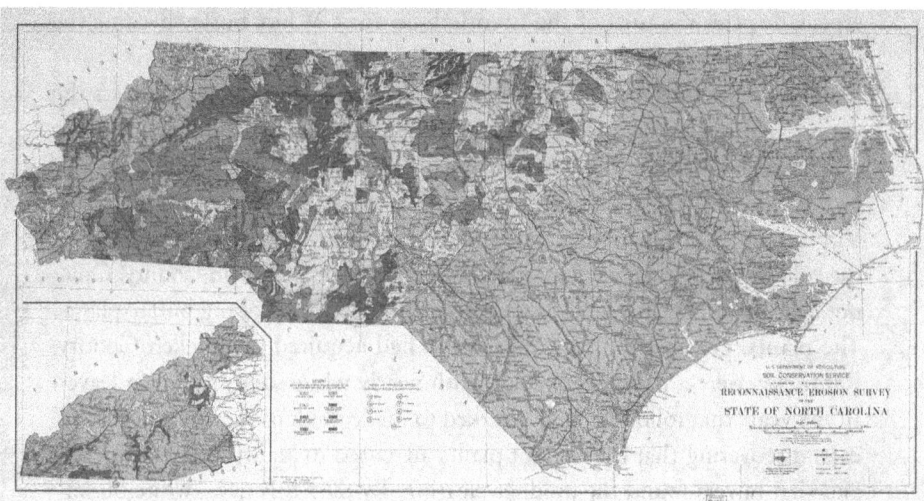

FIGURE 7.2. Both bright leaf and white burley brought mountain farmers environmental challenges, such as eroded fields, highlighted in this Depression-era government study of the state's soil erosion problem. Soil Conservation Service, *Reconnaissance Erosion Survey of the State of North Carolina* (Washington, D.C.: GPO, 1934). Courtesy of the Library of Congress.

market. As a result, sales prices tended to decline and the lucrative figures of the crop's early years were gone.[37]

Farmer independence and the economic benefits it sometimes conveyed were also challenged by changes in the national tobacco industry. Durham, North Carolina, manufacturer James Buchanan Duke used new mechanical cigarette-rolling machines to advance his W. Duke, Sons and Company to the forefront of cigarette sales, and he then used his economic clout to consolidate control over his chief rivals. In 1890, the resulting mergers birthed the American Tobacco Company (ATC), and Duke set about dominating U.S. tobacco. The ATC worked to consolidate first chewing tobacco and then cigarette companies, quickly creating a near monopsony: farmers trying to sell their tobacco in warehouses from Kentucky to South Carolina often found only ATC buyers making bids. The ATC's market dominance led to ever lower prices and fewer options, further reducing bright leaf's appeal on the edges of its cultivation, including southern Appalachia.[38]

Erosion and lower prices for mountain-grown yellow tobacco did not erase the crop from the region. Some farmers continued to raise bright leaf well into the twentieth century, even though crop sizes declined. Others still raised the traditional dark tobacco, cured with old-fashioned wood fires. Increasingly, however, a new variety of tobacco made inroads on Appalachian farms, especially after the turn of the twentieth century. White burley became the iconic mountain tobacco.

Like bright tobacco, white burley has its own origin story that casts the variety as providential and rooted in a particular place and individual. In this case the story is set on the other side of the mountains, in the limestone soils of the Ohio Valley. In 1863, as the Civil War curtailed tobacco production in Virginia and North Carolina and as Kentucky and Ohio growers rushed to fill the void in the market, a tenant named George Webb in Brown County, Ohio, noticed that one of his fields was dotted with peculiar-looking burley plants. The plants, grown from seed that Webb had acquired in Bracken County, Kentucky, were a pale, sickly color, with nearly white stems and thin leaves. In the years that followed Webb worked to make a silk purse from these sows' ears, discovering that the mutant plants air-cured to an attractive light color and that buyers found the product pleasing. By 1867 this new "white burley" had become a sensation on the Cincinnati market and was recognized as a distinct type.[39]

As with any good origin myth, historians contest the details of this seemingly straightforward story. Dates for white burley's first appearance in fields and markets differ, with agricultural historian Lewis Cecil Gray dating light-colored burley cultivation in Kentucky to as early as 1838. Other authors documented experiments with yellow leaf in northeastern Ohio in the early years of nineteenth century, on parent stock that might or might not be considered burley. Farmers besides George Webb sometimes receive credit for the initial discovery (including a co-tenant, Joseph Fore, working on the same farm as Webb), and the name of the landowner appears in different forms too. And exactly when the market embraced the new variety as a recognizable type is also fuzzy.[40]

Despite this uncertain birth story, from the hills of Brown County white burley cultivation quickly spread. Farmers in the rolling counties of southern Ohio and Indiana adopted the crop in increasing numbers, and the Kentucky Bluegrass region rapidly became the center of white burley cultivation. During the 1870s, the commonwealth's tobacco production increased by roughly 70 percent (and Kentucky had supplanted Virginia as the nation's largest tobacco producer thanks to the effects of the Civil War), with the bulk of the expansion in white burley acreage.[41] Although the heart of white burley cultivation in the late nineteenth century lay in the valleys of the Lower Midwest and central Kentucky, the variety also made some inroads on the western edge of the Appalachians, where enterprising farmers sought marketable new cash crops. Farmers throughout the mountains experimented with the suitability of their land for white burley, with significant production developing in the 1880s and

1890s in communities such as Portsmouth, Ohio; the western edges of Kentucky's Cumberland Mountains; and Greene County, Tennessee.[42]

Although in many mountain counties tobacco remained a minor crop, its culture had thoroughly established footholds throughout the region in the decades following the Civil War. By 1873 eastern Kentucky tobacco, though it lagged far behind the leaf production of the Bluegrass and western districts, was promising enough to warrant discussion in Chicago agricultural trade journals.[43] In the 1879 season, farmers from Kentucky's mountainous counties reported a varying degree of reliance on tobacco, from a high of more than 1,500 acres planted in Lewis County on the Ohio River, to just two acres in rugged Harlan County. It is significant that only Perry County reported no acreage in tobacco.[44] The crop's appeal is easy to understand. Although tobacco demanded much hand labor, by 1899 the nation's agricultural census concluded that it "brought to its growers an income per acre of more than five times that derived from all other crops."[45] And white burley seemed to be the most lucrative of Appalachian tobacco varieties, especially as the appeal of bright leaf waned.

Despite the speed with which the burley form spread to new farms, growers insisted that its culture had specific environmental limits. Much as was the case with bright leaf, they stressed that certain soils were crucial to producing quality leaf. Burley "experts" described the best fields as well-drained hillside land with a basic (high-pH) subsoil. As one Ohio farmer wrote in 1875, the palest burley grew on "strong, black, course [sic] river-hill land, and under-laid with limestone."[46] Other ecological characteristics represented promising burley land: for example, certain trees—American ash, beech, buckeye, hickory, basswood, oak, and sugar maple—indicated appropriate soils.[47] As with bright leaf, the literature asserted that particular tobacco forms and particular environmental conditions were inseparable; to produce one, growers had to understand the other.

Just as important as the nature of land that would grow good burley was the nature of the plant itself, and in some ways the phenology and physiology of white burley seemed providential. The white variety proved relatively hardy in the field and in the curing barn, and it averaged more pounds per acre than older, darker forms of burley. Growers could cut the plant whole rather than harvesting its leaves individually—as was starting to become common practice with bright leaf—saving labor, and its air-curing method involved no fuel and less labor than fire- or flue-curing. As a result, white burley growers who raised a variation of the same crop as tobacco farmers in other parts of Kentucky and Tennessee could expect to "receive from two to three times the remuneration for labor expended," a strong incentive indeed.[48]

Manufacturers valued different qualities of white burley. If farmers cured it properly, the leaf was quite durable, proving resistant to mold and rot. Although it tasted fairly bitter, its low sugar content and spongy, absorbent texture meant that white burley accepted flavorings and sweeteners well. These qualities made it adaptable for manufacturing purposes, and the type became an important ingredient in plugs and, by the turn of the twentieth century, newly fashionable cigarettes.[49] Killebrew, a fan of bright leaf but not burley, marveled at the seeming contradictions between white burley's odd characteristics and its market popularity. He remarked, "This variety has few qualities that old tobacco growers would call good. It is very weak. It is thin. It has hardly gum enough in its composition to make it supple." Perhaps, he speculated, an enervated modern society populated by sales clerks and bankers called for such a bland plant. As a result of its qualities, he asserted, "it is probably the mildest tobacco grown, and is admirably suited on that account for consumption by a large class of persons of weak nerves, to whom the use of stronger tobacco is a positive injury."[50]

The business practices of large national tobacco companies coupled with these specific environmental conditions and physiological qualities to make white burley the most important commercial crop in the southern mountains by World War I. The formation of the ATC in 1890, with mechanized cigarette rollers as its key technological innovation, challenged bright leaf growers, but its popularizing of cigarettes would eventually encourage burley's expansion. Although farmers criticized the ATC for its market manipulation, gobbling up competitors and driving down the sales price of tobacco, Duke's company also stimulated consumer demand for cigarettes—a fairly minor American tobacco product before 1900—by driving down their price and advertising their "virtues." The most popular cigarettes sold by the ATC and the companies created by its 1911 dissolution under the Sherman Antitrust Act eventually proved to be the blended brands, combining bright leaf for a mild sweetness with burley, which acted as an absorbent, neutral base to carry flavorings and additives. The two dominant blended brands—Camels and Lucky Strikes—both touted white burley as part of their proprietary blends. Manufacturers' pursuit of ever more burley to fill their cigarettes and Appalachian farmers' pursuit of new goods to market combined to spread white burley across the region in the early twentieth century.[51]

This growth depended on forces similar to those that had fueled the bright leaf boom. The late nineteenth and early twentieth centuries continued the trend of rapid American tobacco specialization. Growers throughout the nation sought to develop and market their own specific "types" of tobacco tailored to consumer niches. In Ohio's Miami River valley, farmers grew seed leaf used in cigars; Louisiana growers produced a rich and dark form known as perique, which became a part of some plugs; Connecticut farmers tended small fields of shade tobacco destined to become cigar wrappers; while in Virginia's and North Carolina's Piedmont counties, bright tobacco growers continued their expansion, undermining the economic position of Appalachian bright leaf farms. White burley was waiting in the wings, however, steadily spreading into the mountains from the west and becoming associated with the region in consumers' minds. Many mountain soils that were too nitrogen-rich for good bright leaf worked well for growing white burley. And the variety's lower labor requirements and air-curing attracted small mountain farmers who had difficulty keeping up with the rapidly changing technology of bright tobacco curing. In farm communities that only a few years earlier seemed destined to grow bright leaf, white burley now seemed the "natural" crop.[52]

Not everyone has agreed that Appalachian tobacco cultivation was a commercial endeavor. Some writers associated with the literary "discovery" of Appalachia wove tobacco into their narratives of isolated, subsistence communities. Geographer Ellen Semple, for example, wrote in 1910 that farmers who grew tobacco in eastern Kentucky did so almost exclusively for household use. And an anonymous author in the popular magazine *Forest and Stream* made a similar claim about tobacco cultivation across the entire southern mountains.[53] In addition to serving these authors' arguments about the Rip Van Winkle quality of Appalachian life (the *Forest and Stream* writer also claimed that mountain men still hunted small game "with old English short bows"), these claims flew in the face of the region's earlier bright tobacco boom and its mounting interest in white burley as a specialized commercial crop.[54] Other more realistic Appalachian critics recognized tobacco as an intrinsically market-oriented crop. Kentuckian Jonathan Gilmer Speed, who chastised the state's mountaineers as backward, violent, uncouth, and primitive, bringing "undeserved reproach upon one of the most peaceful and self-respecting people that ever established a commonwealth," asserted that regional farmers grew tobacco with an eye toward engaging the broader economy. He wrote that the crop was one of the few goods bringing in "such money as finds its way to these people."[55]

Between 1900 and 1929, tobacco was on the march across Appalachia. In the first two decades of the century, bright leaf still grew along the Virginia and North Carolina Blue Ridge, while white burley lured farmers in the westernmost valleys of both states, as well as in East Tennessee and the western edge of the Kentucky mountains.[56] The crop even made footholds in West Virginia, one of the few corners of southern Appalachia with little history of raising tobacco. By the 1910s most of the state's leaf grew in valley farms near the Kentucky and Ohio borders, and farmers sold approximately half a million dollars of white burley each year. State legislators believed tobacco prospects promising enough to authorize research funds and an experiment station with a resident tobacco expert.[57] More intangible but no less important, many farm families came to view burley sales as a safety margin for their farm operations. As one historian who grew up on a small tobacco farm in East Tennessee noted, in the 1920s "it seemed that one or two good burley tobacco crops would easily pay off the $500 mortgage my father had obtained to build the existing house and barn."[58] White burley promised a path to economic stability and even prosperity.

A Bureau of Agricultural Economics (BAE) poll in the early 1930s found 17,805 Appalachian farms that focused on growing "specialty crops," a category dominated by tobacco, and roughly three times as many farms that grew at least some tobacco. Even on the smallest plots, BAE authors observed, the crop was rarely raised just for home consumption. In total the agency estimated that one in seven Appalachian farmers cultivated tobacco and that "receipts from tobacco were frequently an important part of the cash sales, especially on the farms with a small-sized business." Just one in five Appalachian counties lacked an appreciable tobacco crop. This pervasive Appalachian tobacco cultivation had fostered an equally widespread leaf-marketing network. Farmers predominately hauled white burley to warehouses in the towns of East Tennessee, Kentucky's Bluegrass, and Asheville, North Carolina. The remaining bright leaf growers took their crops east, to Danville, Virginia, or Winston-Salem, North Carolina. And the growers of dark-fired tobacco hauled most of their produce to Lynchburg or Bedford in Virginia's Blue Ridge.[59]

The economic crisis of the Great Depression hit rural farm communities with particular force and threatened all Appalachian tobacco growers. Tobacco prices, like those of many commodity crops, tumbled with the stock market collapse, quickly plummeting to roughly half of pre-crash levels.[60] Unlike many "luxury" goods, however, tobacco had some market insulation: addiction ensured that consumers still purchased cigarettes and chewing tobacco in large quantities, even as disposable income declined across the nation.[61]

The most famous federal response to the Depression's challenge to commodity agriculture came in the form of the Agricultural Adjustment Act (AAA) of 1933. The act identified tobacco as one of seven commodities in need of government regulation to stabilize prices and was aimed at balancing crop supply and demand. As a target, the AAA tobacco program set the sale price benchmark to a "parity" figure derived from the averages of sales between 1919 and 1929. This federal pledge to guarantee sale prices at this level was of tremendous benefit to burley farmers in particular, as the variety had sold quite well during the target decade. The program also attempted to alleviate old farmer complaints about corruption in tobacco grading—work often historically done by private warehouse employees—placing grading in the hands of federal agents.[62]

The promised price supports were contingent upon reduced production. The AAA set acreage limits on bright leaf and burley farms, under the theory that reduced supply would inflate market prices close to or above the guaranteed price floor and reduce the cost of the program. Officials created acreage allotments based on the amount of tobacco that farmers had grown during 1919–29 and tied them to specific farms. The tobacco program was voluntary. It required yearly approval from two-thirds of all growers of a particular variety to be in effect, but, once approved, it instituted economic penalties for farmers who violated the acreage limitations. Growers who planted more acres than their allotment faced proscriptive taxes and were not guaranteed parity prices.[63]

The program rules affected farmers and their farms in immediate ways. The linkage of acreage allotments to specific farms froze the expansion of tobacco in place, instantly differentiating contemporary tobacco farms from those without the right to raise the crop, which led to differential values for agricultural land in tobacco regions. The program also often harmed sharecroppers and tenants, as the right to raise tobacco went with the land rather than the grower. For instance, a tenant who had raised tobacco during the 1920s established a tobacco acreage quota for the landowner, but the right to raise tobacco would not follow him if he decided to rent another farm. And when the tobacco program reduced acreage to increase prices, landowners often responded by evicting a portion of their tenants and sharecroppers. As a consequence, anthropologist Peter Benson has declared, the tobacco program effectively made the landowner "the gatekeeper of tobacco agriculture, and stymie[d] the possibility of accumulation or mobility for a tenant or yeoman family."[64]

Despite its controversial aspects, landowners overwhelmingly supported the AAA's program. Parity pricing made growing burley more predictable, and many growers did not mind producing smaller crops of tobacco for the same (or greater) profits. In Kentucky, for example, tobacco production decreased by more than 25 percent following institution of the tobacco program, but farmers realized stronger prices for the leaf they did sell. The Supreme Court struck down portions of the AAA in 1936, but a revised version in 1938 resumed the tobacco program. In the years that followed, the government slightly restructured the regulations—after World War II a cooperative association took over management of the program, determining allotments (which eventually moved from basis in acreage to a "quota" in pounds to reflect increases in productivity per acre), and farmers could sell quota within a given county—

but support for the system of limited production and stable prices remained strong.[65] The program was so popular with small mountain farmers that some new growers entered burley cultivation in the 1940s by exploiting a loophole in the program, essentially surrendering their first year's crop in tax payments in order to establish an acreage basis and allotment.[66] The system was not perfect, and farmers often moaned when officials cut acreage. Sarah Barton, whose family raised tobacco in the early years of the program, recalled the hard times that came when steady reductions "kept whittling down our acre until they got us down to a half acre."[67] But collectively its popularity was demonstrated by the fact that more than two-thirds of Kentucky's burley farmers voted to renew the program every year for seven decades, and even in West Virginia, the Appalachian state with the least historic tobacco cultivation, farmers in its fifteen producing counties consistently and voluntarily participated in the program.[68]

There is debate over the long-term effects of the New Deal's tobacco programs on Appalachian burley farmers. Historian Tom Lee argues that the acreage curtailment cut short the growth of mountain tobacco farming, which was in the midst of an impressive expansion. Without the regulations linking burley to particular farms, he posits, the crop would likely have spread to more and more communities.[69] Most scholars, however, note the ways in which the New Deal's tobacco legislation secured and supported mountain tobacco growers. Stable prices and limited entrance to tobacco cultivation kept existing burley farmers in business at a high rate, and they made even small farms viable. In essence, the tobacco program froze in place what had historically been a fierce competition between regions to produce new varieties. As had been the case with bright leaf, mountain burley production might have swelled only to quickly fade away in the face of competition from cultivators from another region, had the tobacco program's acreage allotments not cemented varietal production to its boundaries in the 1920s, the very time when the crop had made extensive inroads in southern Appalachia.[70]

The New Deal's agricultural activism also brought other programs to aid regional tobacco farmers and secure their crop's footholds in the mountains. Cultivation research and advice flowed from BAE employees, the Farm Security Administration provided loans (even as it sought to relocate some farmers from "marginal" lands), and the Soil Conservation Service helped landowners with plans to slow erosion and maintain soil fertility. Young men working for the Civilian Conservation Corps supported these plans, in many instances planting trees and building erosion-check dams and terracing sloping fields on private as well as public lands.[71]

Just as New Deal programs altered the economics of tobacco farming and manipulated specific landscapes, they also worked to transform ideas about

FIGURE 7.3. "Mountaineers Near Jackson, Kentucky, Cutting Their Tobacco." This image comes from the Farm Security Administration, a New Deal agency that worked to document American rural life. Photo by Marion Post Wolcott, 1940. Courtesy of the Library of Congress.

the crop and its place in the environment. Farmers had long recognized that tobacco was both a demanding and a plastic plant: it required specific soil and climatic conditions to mature into certain forms, but these varying "types" had their own valuable qualities. Put another way, tobacco as a crop had historically been understood as part of a *terroir* (to use a modern term), but that *terroir* had hardly been fixed in place. It evolved over time, and with it growers' ideas about environments changed. The federal tobacco program fixed this moving tobacco landscape in place. It declared certain districts to be tobacco country, while other places were not. As tobacco grew in some counties and not others year after year, this culture-in-place became a de facto stamp of government approval for the best agricultural use of particular soils and farms, conveying the idea that distinctly cultural conceptions rooted in environments were entirely natural. A historic environmental fluidity became a fixed practice.[72]

The legacy of federal intervention in Appalachian tobacco lived on throughout the remainder of the twentieth century. Relatively stable prices, the commodification of the right to raise tobacco through the quota system linked to particular places, and the curtailment of geographical competition all worked to link crop and region, making small burley farms tucked into mountain valleys seem a timeless part of the landscape. Indeed, as the years passed, tobacco became even more important for Appalachian farmers. The relentless expansion of agribusiness consolidated agricultural production across other commodity crops in fewer and fewer hands after World War II, but the New Deal's legacy prevented a similar occurrence in mountain tobacco farms, where small producers "remained financially viable over a period when comparable farms in other regions of the country . . . suffered economic devastation."[73] For small Appalachian farmers at least, the second half of the twentieth century became tobacco's golden age.

&

Tobacco's history reveals the tensions inherent in interpreting Appalachia's agricultural past. On the one hand, the crop proved to be the economic backbone of small farms in a region associated more than any other with subsistence agriculture. At the same time, it was a crop farmers raised for international markets, from its first years growing in Jamestown's fields. Despite the assertion of observers like Semple, mountain farmers always raised tobacco with the intention of selling it for cash, even if that cash was used to maintain independence in other facets of rural life. Scholars who assert the small-scale, locally focused nature of Appalachian agriculture are not wrong, but farmers' inward gaze was hardly absolute. As late as 1929, historian Sara Gregg observes, farmers in Virginia's Blue Ridge were more focused on subsistence than their counterparts in other regions of the country.[74] At that same historical moment, however, nearby tobacco growers were about to embrace a federal tobacco program that would work to regulate production and prices across national markets. They understood their farms as inextricably linked to distant spaces.

These tobacco stories also demonstrate the complexity of people's ideas connecting crops and environments in the region. Tobacco growers of all stripes—dark-fired, bright leaf, or white burley—were convinced that local mountain environments, especially local soils, mattered for growing quality tobacco. These environmental readings shaped where the crop spread, how it was cultivated and cured, and the way it was marketed. In turn these types conveyed ideas about Appalachian environments to national and even international manufacturers and consumers. As tobacco cultivation covered more and more acres, suffered from greater market instability, and intersected with the growing power of the state, the federal government invested mountain environments with specific meanings too. Some programs defined poor farms as "marginal" lands, while other acreage became identified with tobacco cultivation, deemed suitable for participation in global commerce. Over time, people and institutions came to see these divisions as natural, though they in fact emerged at the intersection of actual environmental conditions and cultural constructions. And, in an ironic turn of events, these place- and practice-specific tobacco varieties became commodities on distant markets, treated as standard units of trade, like so many Appalachian natural resources.

Appalachian tobacco's future prospects seem poor, as the end of federal production and price supports with the 2004 TTPP continue to shift cultivation of the weed to other parts of the United States, where economies of scale and labor benefit agribusiness bottom lines. Some regional residents praised the legislation, arguing that it was long past time that Appalachia broke its agricultural dependency on a crop so destructive of human health. Others remained less certain of the long-term effects, noting that it seemed "the beginning of the end for family farming" in places like eastern Kentucky. Private and public programs emerged to help farmers experiment with alternative cash crops and connect with new markets, some of them funded with tobacco company money from the MSA, but no new savior crop has yet emerged to take tobacco's place.[75] What life will look like on former tobacco farms twenty years from today is uncertain. It seems clear, however, that Appalachian farms will be connected to the broader agricultural world, as they have been for so long.

CHAPTER 8

Power

Building an Atomic Appalachia in East Tennessee

ACCORDING TO MYTH, mountaineer John Hendrix was something of a prophet. He liked to hunt and roam the hardwood ridges and bottoms near Oak Ridge, Tennessee, around the turn of the twentieth century, and sometimes he would return from these excursions with visions of what the future would hold. In 1900, give or take a year, Hendrix spent a biblical forty days and forty nights in the woods and returned with an amazing prediction about Oak Ridge's future. He told all who would listen that the sleepy mountain community would one day be a substantial city connected via rails to the broader world. This new Appalachian metropolis would house a massive military factory, and this facility, Hendrix supposedly stated, would produce weapons crucial in "winning the greatest war that ever will be." If there is any truth to this story, it is almost certain that Oak Ridge residents would have chuckled at Hendrix's preposterous claims.[1]

A little over four decades later, events across the globe provided evidence that Hendrix's vision had manifested. On the morning of August 6, 1945, a blinding flash lit the sky above Hiroshima, Japan, unleashing the energy of the sun and marking the first time a nuclear weapon had been used in war. While the Little Boy bomb revealed human mastery of a new and potent form of energy-producing technology, it was also in part a transformation of an ancient source of power much less heralded, Appalachian rain.[2] Although Hiroshima and rural Appalachia are rarely linked, the watersheds of the southern mountains and their tremendous potential energy combined with an influx of federal funding to the Tennessee Valley Authority (TVA) and the Oak Ridge military complex to spur the genesis of the atomic age. The TVA's Fontana and Norris dams captured the regular mountain rains and their turbines converted the water's fall into electricity, a tax-funded power grid transmitted this electricity to federal facilities in East Tennessee, and this abundant, cheap power fueled research into, and eventually the enrichment of, uranium. Although

Hiroshima bore by far the brunt of this revolution, Appalachian nature would not emerge unscathed.

In many ways the story of Oak Ridge symbolizes the transformation of multiple sources of "power" into mountain commodities. The site itself was fueled by hydropower, made abundant by TVA projects initiated in the decade prior to World War II and made feasible by the mechanical energy generated by water moving downhill across southern Appalachian topography. The work of scientists and technicians in Oak Ridge facilities produced a fundamentally new form of power, one that split atoms, with the promise of a new scale of destruction or creation. And a force often underappreciated in Appalachian historiography underlay this production (and consumption) of power: the flow of federal dollars into the southern mountains. Government largesse—for TVA work, national park creation, military installations, and then multipurpose agencies like the Appalachian Regional Commission—fundamentally remade Appalachian landscapes and societies during the New Deal, World War II, and afterward.

Oak Ridge represented this important mid-century Appalachian phenomenon. It symbolized the power of government bureaucracies and federal spending to reshape the southern mountains. Government money, planners, and employees forced rapid change on substantial swaths of Appalachia, sometimes in response to the needs of local people and sometimes in opposition to their desires. As historian John Alexander Williams has noted, as the Cold War drew to a close in the late twentieth century, the military-industrial complex in the southern mountains had become a regional engine of progress and development, "with vastly larger resources and carrying a greater impact than the TVA, the ARC [Appalachian Regional Commission], and urban renewal combined."[3] New military facilities covered mountain landscapes from Radford, Virginia, to Oak Ridge, Tennessee, and Huntsville and Anniston, Alabama. By World War II and the creation of Oak Ridge, no single player wielded more power and influence in any corner of the region than the federal government. Rather than existing as an anomalously isolated region, portions of the southern mountains had been drawn into the central purposes of the state.

In many ways Oak Ridge was and remains an exceptional place. It was one of a handful of Manhattan Project sites and certainly among the two or three most important ones. (The federal government spent more money at the Oak Ridge site than at Los Alamos and the Hanford Works in Washington State combined.)[4] It grew to enormous size almost overnight, transforming Appalachian woods and fields into a complicated network of industrial facilities, military buildings, and residential neighborhoods. Oak Ridge remained something of a secret city throughout the war and is still shrouded in rumor and mystery today, but it was also as representative as it was exceptional: it epitomized the new nature of governmental and economic interconnections in the mid-twentieth-century mountains.

စာ

The Manhattan Project sought to harness nature's power in a new way, and in order to do so it needed specific sites suited to the project's various endeavors, ones with particular environmental qualities if the undertaking was to be successful. East Tennessee emerged as a strong candidate for a site to produce enriched uranium for a combination of reasons. The region was home to a growing cluster of new industries that were proving important in the war effort, most of which relied on the abundant power provided by the TVA. The largest was the Aluminum Company of America, popularly known as Alcoa. The Oak Ridge site was also located just twenty-five miles from TVA headquarters and sixteen miles from the agency's massive Norris Dam, making for easy access to the enormous quantities of electricity the project needed. Indeed, the TVA power grid served as "both a resource and a precedent." The region's hydroelectric potential came from its fundamental geographic and climatic characteristics. One forester notes that "nowhere on the continent, except possibly on the west side of the summits of the Olympic and the Sierra Nevada mountains, is water so plentiful and so pure as in the Southern Appalachians." The upper reaches of the Coweeta watershed in western North Carolina receive roughly 120 inches of precipitation a year, for example, and all that water moving down relatively steep slopes had attracted TVA engineers.[5]

Indeed, the "nature" that attracted Oak Ridge planners was also what environmental historians might call a "second nature," a landscape to a large extent shaped by TVA actions and planning. Natural rivers and their energy had been altered, stored, and released by the agency in ways that influenced planning and development. And the TVA's mission encouraged thinking of Appalachian environmental resources as malleable for human benefit on a scale that few people had imagined before. The agency was in many respects the epitome of the New Deal planning state, working to simultaneously accomplish a number of goals. Agency officials sought flood control, tried to improve river navigation, worked to stimulate economic development, advocated for rural health care and education, and emphasized the transformative power of hydroelectricity. TVA leaders like the agency's first chairman, Arthur Morgan, believed these seemingly disparate efforts were actually of a piece, working in synergy to better the lives of the largely rural population within the boundaries of its authority. For farm productivity, for example, agency dams might cut down on agricultural losses to spring floods. Young men employed by the Civilian Conservation Corps, in partnership with the TVA, would engage in outdoor construction projects such as filling gullies and planting grass strips to help check soil erosion. Agricultural education work could introduce new, more productive crops and livestock, and factories powered by newly abundant hydroelectric energy would manufacture synthetic fertilizers that increased crop yields.[6]

Although there was a large portion of the mountains that the TVA did not physically alter, the agency marked a watershed of sorts in federal thinking about regional nature. The agency's mission was phenomenally ambitious: to better conditions for all people in the Tennessee River Valley and much of its tributary watersheds. Central to TVA leaders' understanding of the region was the notion that "poor land made poor people." Embracing this environmentally deterministic belief, agency officials saw vast environmental planning and manipulation as both a federal right and an obligation. This TVA model became a national approach to the region, shaping a range of government programs in Appalachia, from operation of the eastern national parks to economic development programs like the Appalachian Regional Commission.[7]

To government engineers and military officials alike, the Appalachian mountains themselves offered other appealing qualities besides the water and electrical networks forged by the TVA. Steep ridges separating valleys promised some compartmentalization of a sprawling complex, an important consideration for a site that might some day be the target of air strikes. The relative isolation of Appalachia also promised a measure of ground-level security. The eventual Oak Ridge site, even though nestled in the mountains, was not as rugged as the Great Smoky Mountains to the east or the deeply cut Cumberland Plateau to the west, and it contained a number of potential building tracts. All things considered, Oak Ridge appealed to military planners as a nearly perfect location for a facility that "had to be safe from air attack, not too close to large centers of population, large enough to accommodate four separate plants with flat building areas separated by natural barriers, accessible to rail and motor transport, and on land of reasonable value adjacent to a dependable water source."[8]

Government planners chose East Tennessee because it met this very specific set of criteria, and when it came time to obtain acreage for the project they sought a great deal of it. A huge expanse of land seemed necessary to maintain secrecy and provide space for worker and soldier housing as well as the sprawling industrial complexes necessary to enrich uranium. The eventual target tract spanned substantial portions of Roane and Anderson Counties. Nearly sixty thousand acres, it covered approximately ninety-two square miles.[9] The land was privately owned, and much of it lay in either forest or small, diversified, and relatively cash-poor farms. Much of this farmland was good enough that the TVA had targeted its farmers as recipients of the agency's various agricultural improvement programs, and some agency officials protested that there were more appropriate, less valuable tracts on which to locate the facility. Some of the landowners were willing and even eager to sell, while others strongly resisted surrendering their farms.[10] To speed acquisition of this varied landscape, federal officials turned to eminent domain to obtain the land they believed crucial to the war effort.

By the 1940s, eminent domain had become a common federal tactic to obtain Appalachian land. The regional government relief and conservation programs that sought to alleviate the Great Depression had relied heavily on land purchase—and sometimes forcible acquisition—to reshape society. For example, the creation of the eastern national parks had come largely at the expense of Appalachian landowners. The National Park Service (NPS) in conjunction with Virginia, Tennessee, and North Carolina established Great Smoky Mountains National Park and Shenandoah National Park on lands that were once in private hands. The use of federal land purchases and eminent domain for the parks often drew on stereotypes of the backwardness of Appalachian people and the isolation and rugged nature of regional land.[11] Officials and boosters portrayed those living in the soon-to-be-park grounds as stereotypical hillbillies—poor, illiterate, intemperate, and backward—and claimed removal would improve their lot. Arno Cammerer, assistant director of the NPS, stated, "There is no person so canny as certain types of mountaineers, and none so disreputable."[12] State and federal officials believed these landscapes to be ill-used—or "sub-marginal" in the bureaucratic language of the time—and they argued that they would better benefit the nation's citizenry as parks than as farmland and privately owned forest, but many local people disagreed with those assessments. Nonetheless, state and federal exercise of eminent domain coupled with voluntary sales managed to secure vast swaths of the southern mountains for national parks, and purchase for national forests and state parks would also enclose additional land. Within the national parks, the NPS steadily worked to erase evidence of displaced landowners, razing homesteads, planting trees, and removing roads in an effort to create a wilderness in the image of western parks. The 1930s construction of the Blue Ridge Parkway connecting the two parks also proceeded with the help of widespread land condemnation, furthering many mountain residents' fears of the broad powers of government.[13]

Other efforts to transfer Appalachian land from private owners to public domain during the New Deal also raised regional concerns about governmental power and ambitions. Various New Deal land programs worked to expand federal holdings in the mountains, often by encouraging or pressuring small farmers on relatively unproductive land to sell out and accept resettlement on tracts with better soil elsewhere. In places like Bell County, Kentucky, where once-private land became the Kentucky Ridge State Forest, and on the borders of North Carolina's Pisgah and Nantahala National Forests, where the U.S. Forest Service bought out landowners, numerous transfers took place. These and similar efforts fell under the aegis of several agencies but came to be most closely identified with the fleeting Resettlement Administration (1935–37), directed by Rexford Tugwell.[14] Tugwell declared the administration's mission was to act as a guiding agency "to ease the migrations of people from worn out land and return the land itself to the uses nature would tolerate," and its agents went to work across the Deep South and Great Plains as well as in Appalachia.[15] As noble as Tugwell's goals sounded, some mountaineers were suspicious of the ultimate benefits of government land management schemes. Speaking for many, upon contemplating sale of his land deemed sub-marginal, one eastern Kentucky farmer worried, "I am afeard I would no be satisfied to make a change."[16]

Perhaps the removals of mountain residents most germane to the Oak Ridge efforts were those that occurred at the direction of the TVA. The agency may have aimed to improve farms, land, and lives across the region, but these efforts also entailed a significant amount of destruction and confiscation. The federal government would claim much mountain land for various agency projects. Dams, spillways, service buildings, and access roads would cover hundreds or even thousands of acres of land, and the impoundment lakes that captured flowing water and generated electricity through its release covered hundreds of thousands more. To build these structures and fill these reservoirs (and in some cases to secure economic development zones and water quality buffer strips around them), landowners would have to go, voluntarily or not. Many sold out willingly, happy to receive what seemed like decent prices, while others were forced out by the power of the courts. Various TVA programs offered employment, housing, and educational opportunities to farmers who had lost their land, but some displaced families strongly resented the agency that many Americans viewed as a success.[17]

As was the case with all eminent domain land seizures in the southern mountains, Oak Ridge property owners were of varied opinions concerning the action. According to official estimates, roughly one thousand families were forced from what would become the atomic site, with the government paying an average of $45 per acre for their land, though the actual dispossession total may have been higher: one author questions whether tenants were included in

FIGURE 8.1. The federal government used purchase and eminent domain to take East Tennessee farms like this one for the Manhattan Project. "A Farmhouse along a Road," 1942. Courtesy of the U.S. Department of Energy and the Oak Ridge Public Library.

the figures.[18] In addition to isolated farmsteads, displacement also uprooted a few small villages, such as Wheat, Tennessee. Some displaced people surely saw economic opportunity in the new facility, which promised jobs and paid what some considered fair prices for hardscrabble hill farms. Others seemed resigned, having learned from previous cases—such as Norris Dam and the Great Smokies—that federal authorities invariably won. As Eula Cooper, whose family farm was taken, put it, "It was just something I feel like people, more or less, felt like they had to do."[19]

Although some locals seemed resigned to the land seizures, others were embittered. They did not necessarily question the notion that the land was needed for the war effort, but for one thing they resented the callousness of the supervising officials. Families whose ownership of land went back generations lamented losing those historic connections.[20] And wartime expediency made the difficult situation even more troubling, as families were often given just two weeks from initial condemnation notice to vacate their farms.[21] When the Wheat community was purchased as the future site of one Oak Ridge facility, Bonita Irwin recalled, "They didn't give us any time to pack up and get out hardly. . . . [I] have heard it all these years that they [government employees] burned barns before they could get their hay out. And they tore down some mighty good houses."[22] The process was especially difficult for families that had been through multiple displacements. Eula Cooper held great sympathy for her neighbors who had been forced from their historic home by the Norris Dam project, only to be removed once again by Oak Ridge: "That was more traumatic I think for those people than it was for some of the rest of us." As for Cooper's own family, at the end of their move, when they arrived at their new home, "Mama sat down and cried."[23]

Others saw great opportunity in moving *to* the new facility. The power of Oak Ridge spending lured them in. Ralph Aurin remembered that his father, who had previously read meters for a Knoxville utility company, thought his new Oak Ridge job "grand," and Mary Alexander's father had worked for Alcoa but was enticed away by the new facility's wages.[24] Horace Stanley quit his job with the American Tobacco Company in Durham, North Carolina, heading to East Tennessee to work first in the Holston Ordnance Works (another military complex) before taking a war job at Oak Ridge.[25] As far away as Corbin, Kentucky, schoolteacher Dorothy Gilpatrick also heard "that something big was happening" in East Tennessee and traveled south looking for better work.[26] Likewise, Colleen Black, fresh from her high school graduation in Nashville, moved to Oak Ridge for a position as a leak detector in an enrichment plant.[27] All became part of a flow of humans from across the nation into Roane and Anderson Counties.

Most of these migrants came to work at what was initially designated Site X and then renamed the Clinton Engineer Works (CEW). Within the gates of the CEW were three massive facilities dedicated to producing and experimenting with fuel for nuclear reactions: uranium-235, an isotope that Oak Ridge facilities worked to separate from the more common uranium-238. Because Manhattan Project scientists were uncertain of the technique that would most quickly and efficiently produce uranium-235, Oak Ridge facilities pursued multiple manufacturing techniques. An electromagnetic plant, designated Y-12, sprawled over 825 acres and would prove the most effective in producing bomb material during the war. A gaseous diffusion plant—K-25—housed thousands of additional workers. And a smaller facility, known as Clinton Laboratories or X-10, experimented with producing another nuclear fuel, plutonium, derived from uranium. All three sought to use scientific knowledge, human labor, and an unprecedented investment in machinery to transform one product of nature into another.[28]

New employees at the CEW needed shelter, and an initial transformation of the Oak Ridge environment involved felling oaks and erecting houses as well

FIGURE 8.2. The Oak Ridge facilities, such as K-25 (seen here), employed thousands of workers and scientists. "Aerial View of Oak Ridge Gaseous Diffusion Plant." Photo by Edward Westcott, 1945. Courtesy of the U.S. Department of Energy and the Oak Ridge Public Library.

as factories in their place. The residential districts of the facility became prototypical planned suburbs, presaging sweeping national changes in residential landscapes that would accompany the postwar baby boom. As would be the case elsewhere in the United States in the 1950s, Oak Ridge builders used the techniques of the assembly line to manufacture houses almost instantly. One early resident described the process: "[A] team would come in and set up the foundation. . . . then here would come another crew that would be the flooring. And then the plumbers would plumb. And the electricians would electric. And then the fellows would come in with the sides, then the roof. . . . One day, everything would be finished. . . . Because there was always a family waiting to move in, just as soon as you completed the house." The federal government spent $96 million on the construction of the workers' town alone, including building almost two hundred miles of city streets from scratch. Within a few months, a new house was being completed every thirty minutes.[29] Oak Ridge's residential district also followed the model of mountain planned communities constructed by federal agencies during the New Deal. In the 1930s the TVA had constructed the town of Norris, a model community meant to house, educate, and improve the lives of Norris Dam workers. The Resettlement Administration had also created four Appalachian homestead communities for out-of-work coal miners and factory workers, believing that subsistence farming and craftwork could provide insurance in a faltering industrial economy.[30]

The Coopers, Aurins, Alexanders, Gilpatricks, and thousands of other people from East Tennessee and more distant places swelled the town and plant site. By the end of the war, K-25 would employ twelve thousand workers, and the massive Y-12 complex kept twenty-two thousand people working to produce uranium.[31] Oak Ridge boomed to seventy-five thousand residents, making it the fifth-largest city in the state.[32] The people who moved there in search of work were part of a broader national migration: approximately fifteen million Americans relocated during the war to work in military industries. Most of these people moved from rural places to urban ones, epitomized by the flow of African Americans from the agricultural South to midwestern or coastal manufacturing centers. Oak Ridge was unusual in that it lured people into the mountains rather than out of them. The trend was otherwise. As one historian has noted, during the war, "from rural Appalachia alone, 700,000 people migrated to cities like Dayton, Ohio; Muncie, Indiana; and Detroit, Michigan."[33] For all its futuristic work, planners laid Oak Ridge out along classic industrial city lines, with distinct residential and industrial zones and firm boundaries between the two.[34]

The transformation of landscapes that came with the influx of federal funds and construction presented environmental obstacles, some that no amount of money could immediately address. Government engineers sometimes had a hard time aligning their plans with physical realities. Historian Peter Bacon Hales notes that building Oak Ridge was a process of visceral, material transformation of the site. "Engineers sent in heavy equipment to bulldoze and grade the site, removing such 'obstructions' as trees, vegetation, housing, and the like, and doing their best to convert the topography from hillside to plain. . . . The site was devastated, with much of the topsoil removed and the subsoil already eroding into miniature canyons. Natural drainage systems filled in even as a new drainage system had been started, and the general roadway plan was already staked out, marching up and down the hills."[35] A common result of this constructive and destructive energy was mud. Bottomless, sticky, slippery, omnipresent mud.

Newcomers arriving in the secret city noticed the mud right away, often with surprise or chagrin that such a high-tech site was so obscenely earthbound. Ralph Aurin remembered "stepping off the bus and the mud goes over the top of my galoshes."[36] Ruby Shanks recalled a similar experience as her

FIGURE 8.3. A defining characteristic of wartime Oak Ridge was mud, a by-product of Appalachian nature and continuous construction projects. "Two Construction Workers Work on a Road," 1944. Courtesy of the U.S. Department of Energy and the Oak Ridge Public Library.

excited family arrived, "but when we got out, it was nothing but a mud hole."[37] And Jane Richardson summed up the memories of many early town residents when she declared "Oh yes, the mud. You never forget that. You never forget it."[38] In fact, the modern Oak Ridge Public Library's photographic archives includes "mud" as an identifying tag on a number of images (see fig. 8.3). During dry spells the mud problem became a dust problem. Joanne Gailer, who worked at the K-25 plant, recalled the dust with humor, remembering that she acquired a persistent cough that she believed was triggered by the dust clouds often hanging over the site. The doctor she saw informed her it was a common ailment, shrugging and saying, "Oh, honey, you just have the Oak Ridge croup."[39]

Although it provoked humor on occasion—one poem lamented of the town, "It's crowded. It's muddy. And it ain't pretty"—residents seemed quite self-conscious about the mud.[40] It was as if muddy streets raised questions about whether the Manhattan Project could really unlock nature's mysteries to help win the war. They seemed embarrassed that "people literally left their shoes in the mud sometimes. They would step in the mud and they would pull their foot out and there would be no shoe on it, and they'd just keep going."[41] Evelyn Ellingson remember barefoot walks through the mud to work, where she and her colleagues would wash their feet and slip on shoes to appear respectable on their jobs.[42] And Margene Lyon captures the potential shame of living in the raw town in her recollection of the challenges of maintaining personal dignity when traveling beyond the gates. She "always took an extra pair of shoes, because if the people in Knoxville saw your shoes they knew that you were from Oak Ridge. You either had dust on them or mud."[43]

Town planners worked furiously to transcend the sticky bonds of East Tennessee soil, with mixed success. From the first days of construction, workers built boardwalks above the mud, replacing tracts of forest with miles of horizontal planks. A wooden network of walks would eventually span the town. One former resident marveled that "it had to set some kind of world record for the number of miles that we had of boardwalks in Oak Ridge. But that's the way we avoided mud."[44] Engineers also furiously worked on developing and improving the sewer systems that funneled rainwater and human waste through town, though with less than full success. "Honey wagons" hauling away waste overflows were common sights in the early years.[45] Besides the universal mud, natural processes found other ways to complicate the efforts of Manhattan Project planners. Lieutenant General Leslie Groves recalled one occasion when a mouse got into a plant's vacuum system and temporarily shut down operations. In another instance a bird shorted the entire electrical system of a secure building for several days.[46]

The scientists and other professionals who came to Oak Ridge from elsewhere found both humor and irony in the juxtaposition between the cutting-edge technology and rural rusticity found on-site. They were not above repeating Appalachian stereotypes. Scientist Theodore Rockwell described the town and mountain people condescendingly: "Oak Ridge was a frontier town but a very different frontier town. It was populated with scientists, some Nobel laureates, mixed in with local people, some of whom had never even used indoor plumbing."[47] He also noted what he saw as the outmoded speech and habits of mountaineers.[48] Indeed, Oak Ridge National Laboratory (the overarching name for the production and development facilities) included the story of local mystic John Hendrix as part of its institutional history of the site, employing the folkways of pre–Oak Ridge Appalachia as a humorous contrast to the cutting-edge science that would take place at the federal facility.[49] Perhaps for some participants the rural nature of the compound and its location in the heart of a region long considered isolated and backward, perhaps even dangerous, added to the notion that theirs was heroic work.

The selection of Roane and Anderson Counties for the Manhattan Project had drawn on ideas about geographical barriers and security, and the creation of Oak Ridge worked to make the physical environment even more secure. Planners transformed imagined isolation into real isolation. Oak Ridge was physically enclosed within built as well as natural barriers, and it was guarded by security personnel. Compartmentalization defined the site, with residential and work areas separate, and workers at one job were segregated from those at another. With the creation of the CEW, the government closed formerly public roads, erected fences, posted "No Trespassing" signs, and built observation towers. Individual plants, such as K-25 and Y-12, were enclosed in additional layers of fences and guards. As Peter Bacon Hales has noted, security became "imbedded in every aspect of the physical spaces themselves." In some ways, Oak Ridge came to resemble an especially strict version of a classic Appalachian company town: workers lived in homogenous, contiguous communities; the plants occupied their own tightly regulated grounds; and the firm boundaries of the site helped promote observation, security, and control. In this case the supervising power was the federal government rather than a coal company.[50]

These efforts at control extended beyond bounding tracts of land to labeling and confining human bodies. Mandatory uniforms instantly displayed security clearances for plant workers' access to certain portions of the site. Employees wore color-coded overalls that indicated their fields of work (whether they were electricians, chemists, etc.), and badges numbered one through five attested to their level of clearance. It was thus apparent at a glance if a worker was where he or she was permitted to be. Even speech was tightly regulated. Billboards emphasizing secrecy were erected throughout Oak Ridge. If questioned by outsiders, workers were instructed to "tell them that you're making the lights for lightning bugs or the holes for doughnuts." Oak Ridge managers threatened employees with as much as ten years of imprisonment for speaking about their work outside of the site.[51] In addition to the strictures of security and secrecy, more general southern codes regarding people and their "appropriate" places in the landscape applied to Oak Ridge. The facility was a racially segregated community, following the rest of the Jim Crow South. Worker housing was divided along racial lines throughout World War II. Robert Allen, whose mother worked as a maid in Oak Ridge's residential district, recalled separate softball teams, pool halls, and even card tables for black and white residents.[52]

Although racial segregation seemed to draw little white opposition, the rigidly regulated security landscape had its critics. Within the project itself, some officials thought that Oak Ridge—in particular the residential areas—compared unfavorably with the TVA's nearby planned town of Norris, where, with varying degrees of success, the agency had tried to create a town rooted in local cultural and environmental conditions.[53] Other residents, however, found Oak Ridge's restrictive and controlled atmosphere comforting—it was the ultimate gated community. When the town later contemplated opening its gates to the public, in 1949, some residents initially voted against it. Louise Alspaugh remembers her family and neighbors being reluctant to renew unrestrained interaction with the outside world because they had grown comfortable in the secret city. She recalls, "We felt very safe here. We never locked the door."[54] Within and outside of Oak Ridge a sort of "atomic nostalgia" developed that framed the site and its work as the epitome of the technological prowess and middle-class abundance that characterized popular culture portrayals of mid-century America.[55]

The tight borders of the facility may have promoted a sense of security in some residents, but Oak Ridge also contained a set of hazards connected to the nature of the place. The hurriedly constructed factories and landscapes of heavy equipment and moving machinery were dangerous, and so were the specific processes of separating uranium. The K-25 plant relied on exceptionally toxic gasses flowing through miles of piping, all of it subject to occasional leaks, and the Y-12 plant employed huge and powerful electromagnets that could maim or even kill workers who forgot to divest themselves of metal objects before nearing the equipment. Violence of various sorts was also a part of some Oak Ridge job descriptions. The overarching mission was to build an enormously destructive weapon, one that could end thousands of lives in an instant, and on a more intimate level, security guards on rare occasions shot trespassers or unauthorized personnel. In at least one case, researchers conducted an experiment on an unknowing African American worker, injecting him with a small amount of plutonium to determine its effects on the human body.[56]

These hazards seemed all the more ominous since a veil of secrecy shrouded them. Workers knew that CEW officials and guards took site security seriously and that their work environments were somehow hazardous, although they rarely knew specifics. For example, in an effort to describe her wartime work processing uranium, Earline Banic recounts "separating things . . . one thing from another and we had to put some things with them and shake [them] up through a lot of funnels and then bleed them out whatever they wanted, it would come out. We would take them out and use that and we used a lot of, we had the dark room which we had to dry some of that out." She concludes her vague description, "In a way it was interesting, but you never knew exactly what you were doing."[57] Workers like Banic understood that what they were doing might be dangerous, but they were never provided specific details, only

FIGURE 8.4. Oak Ridge was tightly monitored, surrounded by fences, guard towers, and billboards asserting the need to secrecy. "Guard tower at Y-12," 1943. Courtesy of the U.S. Department of Energy and the Oak Ridge Public Library.

instructions. Likewise, the government obscured the environmental hazards of nuclear work even in its internal records, historian Kate Brown argues, focusing on studies of the movement of radioactive material in abstract ways rather than in careful and systematic analyses of the bodies of workers. As a result, "bodies of exposed people disappeared, dissolved into the heavy physical and mental labor of trying to make insensible isotopes stand up and be counted."[58]

Oak Ridge drew on East Tennessee natural resources and migratory people, and it grew out of mountain landscapes, but just as significantly the project pulled in resources from distant places. For all the secrecy and fences, the site was never truly isolated from the world around it. In fact it was exceptionally dependent on connections to the broader world. Oak Ridge in effect became a giant sink, into which the wealth of the nation temporarily flowed, at great cost. Accounts of Oak Ridge invariably are full of fantastic figures reflecting the site's appetite for natural resources, tabulations of construction supplies, money, and labor required to build and operate its plants. The scale of the construction project—the largest in the history of the Army Corps of Engineers, the supervising agency—inspired awe among locals and also in professional engineers and military officials.[59] A few examples convey the numbers that swelled official reports and histories of the site: In a two-week period of construction, 128 railroad cars of electrical equipment were delivered to the CEW. More than one hundred miles of pipe in the K-25 plant required nickel plating. The electric bill for the first three years of operation of the Y-12 plant was $10 million. The gaseous diffusion facility required 350,000 cubic yards of concrete, its construction would absorb 60,000 railroad carloads of material, and its roof covered forty acres. Construction of the electromagnetic plant used more than thirty-seven million board feet of lumber. The site's steam plant, used primarily as a backup power source, generated twice the maximum electrical output of the TVA's Norris Dam. And such figures ran on and on, for pages.[60]

Oak Ridge National Laboratory's internal history also captures this obsession with the project's immensity in a single celebratory, and nearly breathless, paragraph. In a few short months, the account claimed, "some 3,000 construction workers erected about 150 buildings. The materials list included 30,000 cubic yards of concrete, 4 million board feet of lumber, 4,500 gallons of paint, and 1,716 kegs of nails. Within the boomtown of Oak Ridge itself, a house—sometimes loosely defined—was being constructed every 30 minutes. The bus system in the secret city would be the nation's sixth largest; electricity consumption (largely because of the gargantuan uranium-enrichment plants called Y-12 and K-25) would be 20 percent greater than New York City's."[61] The facilities would draw on the expertise of private corporations as well as the might of the federal government. The heavy machinery manufacturing company Allis-Chalmers designed and built some of the equipment for K-25, as did Union Carbide, which also operated the plant. DuPont Chemical Company constructed the X-10 facility, the University of Chicago helped oversee its experiments with plutonium, and at the end of the war Monsanto Chemical Company assumed its administration.[62]

Some of the materials were less mundane than the concrete, timber, and steel that went into construction. They were precious. Of prime importance was the uranium-238 from which machines and workers in Y-12 and K-25 would separate the much rarer uranium-235 isotopes. The latter would form the fissile material for atomic weapons.[63] The original plant designs called for solid nickel piping throughout the works, but a cautious engineer did the math and determined that following such blueprints would "exceed the entire nickel production of the world," and builders settled on nickel plating.[64] Wartime shortages meant that copper was hard to come by for conductive applications in the plants, and so Oak Ridge builders turned to silver. The U.S. Treasury

Department would ship $300 million worth of silver to East Tennessee for Oak Ridge construction projects.[65] An Oak Ridge resident remembers when the silver arrived because it was heavily guarded. He recalls, "MPs were walking around with submachine guns. . . . We thought probably that the Germans had invaded."[66]

The massive inflow of people and materials, and the secrecy surrounding it, spawned a great deal of speculation and criticism locally and regionally. Rumors swirled about what sort of work was being conducted at the CEW, accompanied by claims that whatever it was, it surely was wasteful of the nation's natural, human, and monetary resources. According to one anonymous observer, it was "a Roosevelt boondoggle merely to have men at work." Stories circulated too of buildings built and then torn down for no apparent reason and construction materials thrown away in mind-boggling quantities.[67]

The purpose of the flow of resources into the secretive confines of Oak Ridge became public in dramatic fashion with the detonation of atomic bombs over Hiroshima and Nagasaki in Japan in early August 1945. For many Oak Ridge residents, this was the first solid evidence of what they had spent months or even years working on. Although some felt ambivalent about what they had helped unleash, most harbored pride at helping end the war. Joanne Gailer recalls news of the first bomb as the "most memorable day" of her life. Even though she worried a bit about the long-term consequences of atomic weapons (so she says upon reflection many years later), she recalls feeling "exhilaration and excitement and celebration . . . people were just very proud of it."[68] Horace Stanley had suspected that the various Oak Ridge facilities had been developing a new sort of explosive, but he did not know of what sort—or suspect its power—until news of Hiroshima. He says of learning of the detonation, "It was a pleasure—you shouldn't be pleased when somebody else gets hurt, I guess, but it was great to know we had done something and it done some good."[69]

The end of the war lifted much of the veil of secrecy obscuring the site, but it also brought a great deal of new uncertainty. Oak Ridge might have withered away once the pressing military emergency had passed. Employment on the site did decline with the war's end, but the quick emergence of the Cold War, which emphasized the atomic might of the United States, ensured continued federal spending on Oak Ridge. After the end of the war, Union Carbide continued to operate K-25 and Y-12. Monsanto—which had administered a Manhattan Project research site in Dayton, Ohio—took control of X-10, which in 1948 became Oak Ridge National Laboratory, dedicated to researching nuclear reactions for the newly formed Atomic Energy Commission. After the attack on Japan, and then the opening of the gates to Oak Ridge's residential areas in 1949, much of the mystery and speculation surrounding the "secret city," evaporated, but the power of the atom and federal dollars seemed destined to remain a part of Appalachian Tennessee.[70]

Oak Ridge transformed environments and human lives, reordering flows of timber, electricity, concrete, precious metals, radioactive ores, and workers, and it proved just as important as a symbol of the new place of the federal government in Appalachia. With the Manhattan Project following hard on the heels of creation of national forests, national parks, and regional New Deal programs, government money had become an integral part of the southern mountain economy. These endeavors did much to reverse traditional economic relationships. Before the 1930s, Appalachian resources tended to flow out of the region, to the benefit of consumers in other parts of the nation or the globe. From the New Deal on, there was also a persistent inflow of government spending. Large quantities of Appalachian timber, coal, and minerals still left the mountains for distant markets, but federal cash steadily infused pockets of the mountains too.

This federal spending took multiple forms. Some development followed closely on the lines of Oak Ridge. Two new munitions plants in the New River Valley of southwestern Virginia—a region with a long history of industrial mining and manufacturing—expanded along with Oak Ridge during World War II. The New River Ordnance Plant at Dublin and the Radford Ordnance Works at Radford both supplied shells for the military effort and grew exponentially during the war. The facilities enjoyed "natural" advantages similar to those of Oak Ridge: relative isolation, abundant water, and cheap electricity. By the end of World War II, the Radford plant alone employed more than nine thousand workers. The two facilities consolidated into one massive plant in the 1960s, and Cold War tensions kept production steady.[71] At the southern tip of the mountain range, Anniston, Alabama, grew into another node of the military-industrial complex. World War II promoted an expansion of the town's Fort McClellan and led to construction of the Anniston Army Depot, which in turn prompted the growth of Monsanto's local ancillary operations. The Anniston Army Depot would go on to become a storage site for the government's chemical weapons.[72] Similar military installations dotted the region.

The 1960s brought a new wave of money to Appalachia. John F. Kennedy's President's Appalachian Regional Commission determined that government funding and oversight were necessary to solve entrenched regional economic problems, and in 1965 Congress passed the Appalachian Regional Development Act, overseen by a new agency, the Appalachian Regional Commission (ARC). Another element of President Lyndon Johnson's War on Poverty, the Economic Opportunity Act, also poured money into Appalachia, spawning hundreds of county, state, and regional programs aimed at stimulating the Appalachian economy or attacking the "culture of poverty." Regional colleges and universities took part in this wave of funding and investment. West Virginia University, the University of Kentucky, the University of Pittsburgh, the University of North Carolina at Chapel Hill, Berea College, and other institutions to a lesser extent conducted studies, housed academic experts, and formed advisory panels on a wide range of Appalachian concerns.[73]

The ARC in particular became a regional political power. Historian Ronald Eller has called the agency's work "a domestic version of the Marshall Plan . . . both an expression of American political culture in the postwar years and a sign of popular confidence in the ability of science and technology to produce the good life." By the 1970s, the ARC's annual budget had grown to approximately $300 million. The agency promoted a specific vision of Appalachian development, one that bypassed the iconic rural landscapes of the mountain South. ARC spending focused on infrastructural development—new highways, rail lines, and bridges—and concentrated on creating mid-sized cities that would, in theory, provide jobs and services and fuel growth in the surrounding countryside. Throughout the 1960s and 1970s, four out of every five dollars the agency spent went to road construction or repair. This focus left the most remote and rural districts further behind than ever.[74]

∽

Oak Ridge was at once unique and representative of Appalachia from World War II on. Like other atomic sites, Oak Ridge was a landscape shaped by a number of unusual factors: an intense desire for secrecy and security, advanced technology, and new and persistent environmental hazards to landscapes and human bodies.[75] But it was also, like an increasing portion of the rest of Appalachia, a landscape to a large degree created and supported by government spending. Oak Ridge became the site of federal investment of an intensity matched in few other places, unique in scale but common in form.

On the surface, this reliance on military spending and high technology seems quite un-Appalachian, or at least fails to fit neatly into stereotypical portrayals of a benighted and backward mountain South. For this reason, the history of Oak Ridge has not often been explained as an "Appalachian" story, especially by Appalachian studies scholars.[76] The Manhattan Project just seems too atypical, as if placed in the region by accident or whim. This disjunction between Oak Ridge reality and popular culture ideas about the highlands may help explain the popularity of the Hendrix story beginning this chapter. As sociologist Lindsey Freeman argues, "It is curious that Oak Ridge, the Atomic City, a place that sees itself as a stronghold of high culture, would trace its origin to a notorious hillbilly; it is perhaps even more curious that a city devoted to science would rely so heavily on a mythic prophecy to tell the story of its beginning." Freeman attests that the Hendrix myth became popular because it obscured the more complicated and sometimes ugly realities of forming Oak Ridge, replacing them with humor. Gone were the land seizures of eminent domain, the razing of forest, the omnipresent mud and muck, and racial segregation. Hendrix's tale leaped the gap between old ideas about Appalachia and the very different present, spanning the chasm between images of fiercely independent Scots-Irish farmers and a town of federal employees working for the defense industry, and conveying the idea of a sort of "atomic manifest destiny."[77]

This story arguably elided more than it revealed, tapping as it did into national ideas about Appalachia and the Manhattan Project. It was in fact the very Appalachian nature of Oak Ridge that attracted military planners in the first place. The region's hydrology, its geography, and its cheap, rural land made it an appealing site for a massive, secure complex. And it set a new norm for the region, becoming part of the story of broader twentieth-century Appalachia. Landscapes once characterized by farming or extracting timber or coal became the sites of new highways, military bases, tourist attractions, vacation homes, reservoirs, scenic parkways, and national parks and forests. As agencies like the ARC formalized a federal stance that Appalachia was different, full of intractable economic and social problems, government spending began to connect the southern mountains to congressional budgets almost as firmly as had been done with the American West. Ideas about the old Appalachia helped forge a new one, one where stereotype and reality were often far removed, facing one another over a mountain of federal dollars.

CHAPTER 9

Coal

Sludge Ponds and Vanishing Mountains

"COAL HAS ALWAYS CURSED the land in which it lies. . . . It mars but never beautifies. It corrupts but never purifies."[1] Harry Caudill, a Kentucky lawyer who became a social and environmental activist and eventually a state legislator, wrote these words in the preface to his poignant 1963 exposé of mountain poverty, *Night Comes to the Cumberlands*, indicting coal and the coal barons for Appalachia's fate. Attacking the coal industry in eastern Kentucky in the 1960s was bold, perhaps even foolhardy, but Caudill was convinced that the relationship between people and nature was out of balance. The region faced an environmental disaster, he claimed, one that lay at the heart of persistent social and economic problems in the mountains. Appalachia was being destroyed by its economic connections, not its isolation. Few events would more strongly support Caudill's arguments for the hazards of coal and demonstrate the ineffectiveness of mining reform efforts than one that took place almost four decades later: a coal sludge spill that occurred in Martin County, Kentucky, in 2000. It was an event that proved the prescience of Caudill's warnings and demonstrated just how little true reform had taken place since the publication of his book.

During the early morning hours of October 11, 2000, a coal slurry storage impoundment, or "sludge pond," belonging to Martin County Coal Corporation (MCCC) failed, releasing more than three hundred million gallons of toxic slurry through abandoned mine shafts that lay under the lake of waste and into Wolf and Rockcastle Creeks. The slurry was as close to solid as liquid, a thick, oozing mass the consistency of lava, laden with magnetite and other heavy metals produced during the coal washing process (see fig. 9.1). It was a disaster of mind-boggling proportions, roughly half again larger than the *Deepwater Horizon* spill that dumped an estimated 186 to 226 million gallons of crude oil into the Gulf of Mexico in 2010. The liquid pollutants flowed downstream, killing fish and other aquatic life and contaminating wells and town water supplies as they went, eventually affecting residents along the branches of the

FIGURE 9.1. Washing coal (removing ash and other impurities in order to produce "clean" coal) produces massive quantities of wastewater, which coal companies store in sludge impoundments like this one on Shumate's Branch, in West Virginia. "View of the Valley Sludge Pond from the Mine Entrance End to the Dam." Photo by Lyntha Scott Eiler, 1995. Courtesy of the Coal River Folklife Collection, American Folklife Center, Library of Congress.

Big Sandy River as the tide of waste flowed toward the Ohio River. In communities like Fort Gay, Kermit, and Kenova, West Virginia, and Inez and Louisa, Kentucky, schools and water treatment plants closed and water had to be piped or trucked in from outside the spill zone. Life for all creatures downstream of the failed impoundment would be affected for years to come.[2]

For all its grave environmental consequences, the Martin County spill could have been worse. The sludge impoundment held well over a billion gallons of coal waste. The breach took place on one side of the eighty-acre pond, where the company could plug and repair it with relative ease, rather than at the lowest point. As a result, approximately 80 percent of the slurry remained in the impoundment, and the force of the breach and the total extent of the pollution were less than they would have been in the case of a complete failure. Although the environmental and economic costs would prove steep, there were no human casualties.[3]

The immediate human costs might have been much graver. In 1972 in Logan County, West Virginia, a similar coal slurry impoundment had catastrophically failed. In the steep terrain the wastewater swept down the valleys below like a thirty-foot tidal wave, giving little warning to people in its path. When rescuers combed the wreckage, they found 126 dead, more than 1,100 injured, and 4,000 people made homeless in one of the nation's most costly environmental calamities. Many survivors were psychologically scarred for life. This Buffalo Creek spill briefly brought national attention to the broader human and environmental costs of coal, as images and reports of the disaster trickled out of West Virginia, but it was just as quickly forgotten. Slurry impoundments remained a standard feature of Appalachian coal mining, part of the cost of doing business, and the impoundment that failed in Martin County nearly thirty years later was in most important respects similar to the flawed one at Buffalo Creek.[4]

There was one key difference, however. Martin County was the scene of a newly intensive form of coal mining that had come to dominate the southern mountains, a method its critics would label mountaintop removal (MTR). Major coal corporations, such as MCCC's parent company, Massey Energy, have touted MTR methods, which they label surface mining, as marvelously efficient techniques. According to this narrative, MTR allows a few equipment operators, engineers, and explosives experts to quickly and cheaply expose and extract entire coal seams. This cheap coal furnishes abundant domestic energy and reduces the nation's dependence on foreign oil. MTR employees are highly paid, and their work is safer than that of underground miners. These operations carefully consider the environmental and cultural costs of their techniques, and coal company officials strive to leave behind landscapes that are healthy and more useful (flatter, anyway) than before mining took place. Industry supporters point to grassy hilltops, new buildings on old mine sites, and replanted trees as evidence of MTR successes.

Mountaintop removal opponents tell a decidedly different story. They argue that MTR is the epitome of environmentally destructive corporate profiteering. Companies avoid the once largely unionized workforce of deep shaft mining, replace humans with machines and explosives, and pass along the environmental costs to neighboring landowners. Dust, runoff, deforestation, flooding, and noise from MTR operations make large tracts of central Appalachia all but uninhabitable, vast ponds and waste pits like the one that failed in Martin County hold the toxic slurry water used to clean coal and threaten anyone who lives downhill from them, and environmental cleanup operations that follow mining fail to truly rehabilitate mine sites. To make these consequences all the more tragic, the mines produce coal that dirties the air, contributes to global warming, and floods an already saturated market, as much of it is exported overseas where it powers competitors to American industry. There is little common ground between the two interpretations.[5]

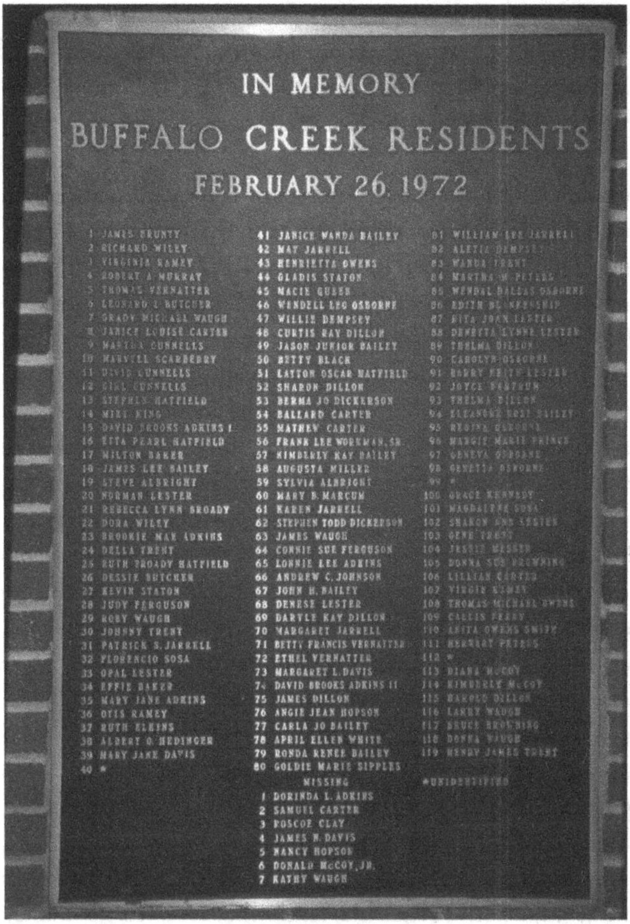

FIGURE 9.2. The failure of a waste impoundment at Buffalo Creek, West Virginia, on February 26, 1972, resulted in the death of 126 people. "Memorial Marker at a Church in Buffalo Creek, Listing the Victims and the Missing Persons of the Buffalo Creek Disaster." Photo by Mary Hufford, 1997. Courtesy of the Coal River Folklife Collection, American Folklife Center, Library of Congress.

The disaster in Martin County was one side effect of MTR, but the spill had historic roots. An explanation of why sludge spread from a breached slurry impoundment down eastern Kentucky watersheds and into the Ohio River requires an understanding of the development of regional strip-mining. Although the massive scale of mountaintop removal is a product of the regulatory environment and technological developments of the last thirty years, strip-mining is an old method of obtaining Appalachian coal. And the story of strip-mining in turn begs a brief foray into the history of the late-nineteenth-century Appalachian coal boom and the legal and social details of regional mineral rights.

The southern mountains had long been the site of real estate speculation, but land agents for coal and timber companies first entered central Appalachia in significant numbers in the late nineteenth century. There to secure supplies of coal and lumber for industrial purposes, these agents sometimes turned to deceit or abuse of the law to acquire land, such as the practice of using sympathetic courts and antiquated and confusing regional property records to contest land titles. If coal agents could force the public auction of tracts, the companies were all but assured of gaining possession since only they had the financial resources to purchase large properties. Some buyers also took advantage of landowners' illiteracy or relied on intentionally deceptive contracts written in complicated legal jargon. Although deceit was common, agents more often made straightforward offers to purchase land and met a warm reception from mountaineers. The reason was simple: they offered to pay good money in cash-starved local economies for poor hill farms that had attracted few previous buyers. Often the outside agents or corporations worked hand-in-hand with locals, especially mountain town lawyers and merchants who understood community relationships, kept an eye out for promising land, and saw economic opportunity in the spread of industry through the mountains.

In many cases agents offered to purchase only the rights to underground minerals, leaving use of the surface to the landowner. In some instances this seemed like a good deal for both parties: the landowner received the money when the deal was struck and retained the use of his or her farm, and the coal company gained the right to mine the coal without having to pay property taxes on the acreage. These agreements that separated ownership of the surface and the subsurface came to be known as "broad form deeds," and their true power became apparent as the feeder branches of railroad lines like the Norfolk and Western and the Chesapeake and Ohio penetrated deeper into Appalachia by the end of the nineteenth century, making large-scale coal mining increasingly possible and profitable. The language of the broad form deeds often granted mineral rights owners the power to extract coal through the most efficient means. In the eyes of the law, this permitted coal companies to do such things as clear timber for constructing access roads and mine entrances, use underground explosives to free the coal, and build mine structures and coal storage facilities as needed on the surface. All of these activities significantly impaired landowners' use of the surface, in the most egregious cases rendering entire farms virtually worthless. Coal operators who mined land they owned and those who operated under broad form deeds enjoyed virtually the same rights and privileges.[6]

Traditional underground coal mining took advantage of the broad form deed, but surface mining would bring new, harsher consequences for property owners and their neighbors. Surface or "strip" mining—exposing the subsurface coal by removing the dirt and rock above it—had a long history, even if it was not the predominant method of obtaining coal until the mid-twentieth century. In the nineteenth century, operators in the Illinois and Indiana coalfields employed strip-mining techniques where coal seams lay relatively close to the surface, using traditional agricultural equipment. Miners plowed the topsoil to loosen it and then removed the dirt with horse-drawn scrapers, repeating these steps until they reached the coal bed, often found just a few feet below the ground level. The introduction of steam-powered equipment made the process more efficient by century's end. During the early 1900s, there was limited strip-mining in Appalachia too, but only where coal seams were unusually accessible.[7]

After World War II, Appalachian strip-mining rapidly expanded. Part of the reason was an increased national appetite for electricity, most of it produced by coal-fired power plants. After nearly two decades of depression and war, Americans were ready to consume their way to the good life, and, along with personal automobiles, electric home heating and air conditioning systems and small appliances were a cornerstone of the new economy.[8] Road-building efforts in Appalachia during the New Deal and World War II also made a new class of small mines (known as "truck" operations) possible, as companies with just a few employees could now truck coal to tipples located on the rail lines. With the high cost of coal during the postwar years and better road networks, these small mines proliferated, adding to the glut of Appalachian coal.

New technologies also increased the efficiency of larger underground mines. In the old tunnel and shaft mines, the continuous miner, a machine that chewed away nonstop at the face of the coal seam, largely did away with the old, labor-intensive, and dangerous techniques of drilling and blasting. Conveyor systems with endless loops of belting replaced mules, carts, and drivers in moving coal from the face of the seam to the mine's mouth. And steel roof bolts took the place of millions of board feet of wooden supports used to prevent ceiling falls, eliminating many timbering and placing jobs as well. A 1950 labor contract between large operators and the United Mine Workers of America union also hastened companies' mechanization, as higher worker salaries and more generous benefits made the new technology more fiscally appealing, especially as increasingly efficient mechanized mining promised to drive the competing truck mines—themselves a product of new technological systems—out of business.[9]

Most significant, mine mechanization eventually led to more surface mining. Better and bigger equipment, the large payments that accompanied it, and more costly labor encouraged large companies to explore or expand strip-mining operations. At the same time, surface mining held growing appeal for smaller operators, as a strip mine required fewer employees and less overhead than an underground mine of equivalent size. An abundance of explosives (often military surplus) and bigger bulldozers made exposing coal more efficient and profitable than ever before. The technique of "high wall mining" grew increasingly common. In this method miners located a coal seam where it broke the surface on a mountainside, dug away the overburden (everything that was not coal) above the seam, and then removed the coal. These strip-mining efforts only extended a certain distance into the slope before removing the ever-thicker overburden became unprofitable. High wall mining left spoil piles below the cut and a hazardous cliff face—the "high wall"—where mining ceased.

Auger mining soon supplemented high wall techniques. Large motorized augers bored into the coal seam beyond the point where removing the overburden was practical. This extended the reach of the high wall methods but made hillsides even more unstable in the process, as augering boreholes honeycombed mountaintops. There was even some small-scale mountaintop removal mining by the 1960s. Where rich coal seams were discovered close enough to a ridge line, miners bulldozed the top of the peak into an adjoining valley to get at the entire seam. By 1963, these combined forms of "strip operations produced nearly one-third of the bituminous coal mined in the United States," much of it coming from central Appalachia.[10] As a consequence of these mechanized operations, even as coal production increased, the number of Appalachian coal miners declined precipitously, from 475,000 in 1945 to just 107,000 by 1970, and unemployment in the coalfields became chronic.[11]

Ironically, the Tennessee Valley Authority (TVA) did much to drive the expansion of central Appalachian strip-mining. Although the TVA was founded as an agency tasked with overseeing conservation initiatives in the mid-South as well as accomplishing its better-known missions of flood control and hydroelectric power production, during the postwar period it increasingly shifted its focus to furnishing abundant, cheap, coal-fired electric power. The TVA's electricity-generating ambitions soon outstripped its hydroelectric capacity, and during the 1950s the agency began building mammoth coal-fired power plants. Appalachian coal fed them, and the cheapest coal of all came from strip mines. As historian Ronald Eller summarizes, by 1960 "the TVA itself had purchased the mining rights to almost 100,000 acres of coal in eastern Kentucky and east Tennessee, and power plants and other consumers across the Midwest had turned to burning cheap, surface-mined fuel."[12] The TVA essentially displaced the costs of modernizing the Tennessee River Valley onto the mountains and coalfield communities of central Appalachia.

Broad form deeds, advancing mining technology, and a national electricity market made Appalachian surface mining profitable, but another critical component in the spread of stripping was the rulings of regional courts, especially in Kentucky. In the past the courts had regularly judged that broad form deeds granted mineral rights owners broad powers. In the legal language of the courts the only action not permitted to the mineral owner was "oppressive, arbitrary, wanton, or malicious conduct" in extracting coal or other resources.[13] Strip-mining opened new possibilities, horrific ones for surface owners. As one southwestern Virginia miner who was also a hunter noted, "I started seeing strip mining [in the] very early '60s. And when I would walk through a strip mine and I see what it did to the forests and the water—the streams—and the wildlife, I just became incensed about how it destroyed the earth."[14] Denuded hillsides and enormous piles of mining spoils also created new drainage challenges in parts of the mountains, as the mined landscapes could no longer absorb as much rainfall. The result was devastating localized flooding, as was the case in the destructive 1977 floods in eastern Kentucky.[15] In essence, strip-mining threatened to completely destroy the surface of both the mine site and adjacent properties in order to access minerals, something not imagined when most broad form deeds were drafted.

Strip-mining seemed a perversion of the old agreements, and so some states prohibited stripping under broad form deeds. But in Kentucky the courts consistently ruled that strip-mining was permitted under the deeds and that it was not "oppressive, arbitrary, wanton, or malicious."[16] By the 1950s, Kentucky courts had ruled that not only could coal operators strip-mine under broad form deeds, they also need not pay damages to the surface owner, even if the entirety of the surface was removed and relocated. One vocal critic, Harry Caudill, declared the commonwealth court system's "decisions medieval in outlook and philosophy" and warned that "this long line of judicial opinions opened the way for what may prove to be the final obliteration of the [Cumberland] plateau's future as a vital part of the nation and its history."[17] In removing the mountainsides, strip-mining threatened to erase regional communities and culture.

Local worries about the effects of strip-mining grew apace with the acreage mined after World War II, but it was Letcher County, Kentucky, lawyer Caudill who brought national attention to Appalachian problems. Caudill had grown increasingly dissatisfied with the socio-economic and environmental plight of eastern Kentucky, and he believed that large mining companies bore a significant share of the blame. He would attack strip-mining with a pen rather than in the courts. Caudill's *Night Comes to the Cumberlands* was part folk history of Appalachia, in the tradition of John C. Campbell and James Watt Raine, part exposé of the mining industry and government corruption, and part plan of action, and the book met national acclaim. Caudill saw plenty of blame to go around for the situation in eastern Kentucky. He blamed state governments for caving in to coal operators, coal operators for their greed and callousness, the federal government for misguided social welfare programs, and Appalachian people themselves for their willingness to accept dependency (even though he was sympathetic to the burdens he believed history had placed on them). Caudill went on to serve three terms in the Kentucky House of Representatives and taught history at the University of Kentucky. He remained a vocal critic of strip-mining, publishing additional activist writings, such as *My Land Is Dying* (1971), and agitating for national legislation to control or prohibit strip-mining.[18]

Some Appalachian residents took Caudill's activism a step further, engaging in direct action to oppose strip-mining. Most of these acts took place in the late 1960s, when social unrest and regional activism on poverty and social justice were at their peak. Kentucky was the hotbed of the movement, where some prominent examples of direct action against mining made national news. In 1965, Knott County landowner Dan Gibson and his neighbors obstructed a mining operation with their bodies, carrying rifles to meet miners. Also in 1965 in Knott County, the elderly widow Ollie Combs and her relatives blocked bulldozers threatening their family land and were arrested for their troubles. In 1967 an activist organization, Appalachian Group to Save the Land and People, met Puritan Coal Company bulldozers on their way to strip-mine the farm of Jink Ray and perhaps used explosives to deter miners, though it was never proven. The same year a pair of diesel shovels were dynamited in Knott County and the "Mountaintop Gun Club" leased properties threatened by strip-mining and turned them into temporary firing ranges in an effort to intimidate miners. In the summer of 1968, anonymous vigilantes blew up three quarters of a million dollars worth of strip-mining machinery in Leslie County, Kentucky. Activists also temporarily shut down two eastern Kentucky coal mines through on-site protests in 1972.[19] More common was peaceful and law-abiding protest, such as Virginian Frank Kilgore's founding of the advocacy organization Virginia Citizens for Better Reclamation and Letcher County resident Raymond Bush's early 1960s petition drive to ban strip-mining in the state, an effort that garnered thousands of signatures.[20]

By the late 1970s Caudill's writing, grassroots protests, and a growing number of other accounts connecting Appalachian poverty, environmental degradation, and industrial exploitation led to new strip-mining legislation. The
Surface Mining Control and Reclamation Act of 1977 (SMCRA), the most important of these regulations, established some overarching guidelines for strip-
mining. On its face the act promised to check the worst excesses of new mining
techniques, but coal industry lobbying and sympathetic politicians had crafted
legislation full of gray areas. As written, the act required that mining companies rehabilitate mine sites. To accomplish this they would have to re-create
the "approximate original contour" of the surface (although at a slightly lower
elevation, reflecting the missing coal), they should leave buffers of undisturbed
land bordering streams, and they should replant the rebuilt slope. If strictly
followed, these guidelines might result in a landscape that was more disturbed
than a clear-cut but certainly less devastated than a leveled spoils pile.[21]

Although the SMCRA was the promising result of activism, in some ways it
ended up representing a victory for coal operators. Despite its new stipulations,
the act did not ban strip-mining outright. In fact it gave it official, if regulated, sanction. What was not included in the act was as important as what
was printed. Historian Chad Montrie summarizes the SMCRA's shortcomings
when noting that it "allowed stripping on steep slopes and alluvial valley floors,
said nothing about coal reserves owned separately from the surface and mined
without surface owner consent (except federally owned land), and permitted
mountaintop removal . . . as well as impoundments for slurry waste."[22] There
proved other gigantic loopholes in the legislation. If operators claimed they
were creating a landscape of higher use—for example, land suitable for livestock pasture, housing, or other development—they might obtain a variance
from having to restore the original slope. Mining companies soon touted the
utility of flat land in a mountainous region—what a perfect site for a golf
course or airport!—and sympathetic government officials were forthcoming
with variances. The terms of revegetation were also imprecise, and standard
practice became planting grass or legumes to "rehabilitate" old strip mine sites,
with little follow-up to ensure the seeding took or lasted past the first year or
two. If companies did plant trees, they were usually white pines and locusts,
which grew quickly in poor soils, rather than mast-bearing hickories and oaks
that might eventually reproduce the preexisting forest composition. As long as
industry could influence the officials overseeing compliance, the regulations
that did exist remained fluid.[23]

If the SMCRA was ripe for exploitation, conditions at the end of the twentieth century created great economic incentive to take advantage of its flaws.
Regular turmoil in the Middle East contributed to uncertainty in petroleum

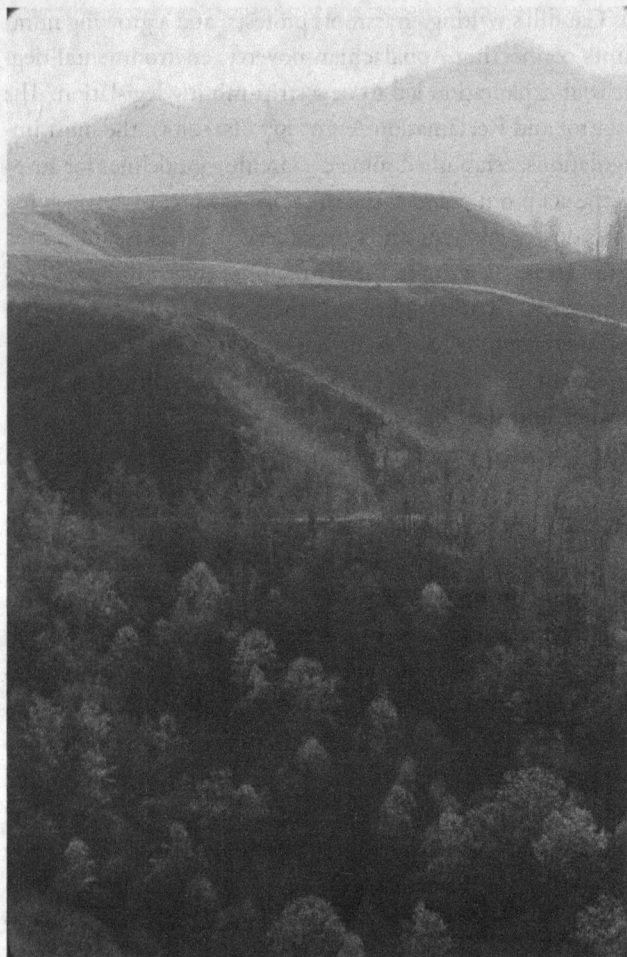

FIGURE 9.3. "Reclamation on the Costain Mountaintop Removal
Site, White Oak [wv]." Photo by Lyntha Scott Eiler, 1997. Courtesy
of the Coal River Folklife Collection, American Folklife Center,
Library of Congress.

supplies, making coal a more appealing energy source. Mining machinery continued to grow larger and more efficient, and labor grew ever more expensive. In 1990, new amendments to the Clean Air Act demanded reduced emissions from the nation's power plants. Although this was a victory for air quality advocates, it proved a great loss for opponents of MTR, as power plant operators turned to the low-sulfur coal produced by Appalachian surface mines to meet the new guidelines. Playing on this mixed environmental message, the mining industry adopted advertising campaigns touting "clean coal," neatly hiding the very dirty process of producing strip-mined coal behind the reduced particulates generated by its combustion. And in 2008, the George W. Bush administration relaxed strip-mining regulations even further. Administration appointees reduced the mandatory buffers between strip-mining sites and streams and rewrote the definition of "mining spoils" in a way that allowed debris from stripping to obliterate miles of mountain streams without being subject to the regulations of the Clean Water Act. New mining permits proliferated, and by 2009 approximately 450 Appalachian mountains had lost their tops to strip mines.[24]

Exactly what sort of environmental crisis was the new form of mining? Critics of MTR often argued that it was a more thorough destruction of nature than almost any other human activity and as such could not be justified solely on economic grounds. As mentioned, the SMRCA had decreed that mining companies had to rehabilitate mine sites, restoring the original contour if possible and revegetating the new land's surface. In practice this typically resulted in a relatively flat tract of land planted in grasses and a few legumes. Although this rehabilitated landscape might look green—pastoral even—for a few years, without regular fertilization of the relatively sterile subsoil that composed the new surface the plant cover soon withered and died. Southern West Virginia resident Jim Foster noted of reclaimed MTR sites, "after a year or so, everything washes away. You don't have anything but just a desolate wasteland, just like the moon surface or something."[25] Kentucky farmer, essayist, and poet Wendell Berry suggests that the very idea of rehabilitation is dubious, asserting that "no strip-mined land, however regulated and reclaimed, is as good as it was before—and, in human time, it is not going to be as good." Compounding the difficulties of reclamation, no matter what its intentions its labor "is still done by people who work for absentee owners and absentee governments, and who will not have to live with the consequences of their work."[26] Opponents of MTR also derided industry claims that rehabilitated landscapes would provide valuable, relatively flat land for development. Would former mine sites in fact become airports, shopping centers, or golf courses? Jack Spadaro, who was one of the federal officials who investigated the Martin County spill, assessed such claims: "Well, about a million and a half acres have been mined in this area. And a very, very small percentage—less than 1 percent—of the surface area has actually been used for alternative postmining land use. . . . That whole alternative postmining land use is simply bullshit."[27]

In addition to containing the loophole that allowed mining companies to superficially "rehabilitate" landscapes, the SMCRA also permitted the creation of massive coal slurry impoundments like the one that failed in Martin County. Operators cleaned coal using millions of gallons of water before shipping it

FIGURE 9.4. Mountaintop removal mining entails the removal of all soil and rock ("overburden") covering a coal seam, as seen at this mine in southern West Virginia. "View of Arch Mineral Corporation's Mountaintop Removal and Reclamation Project, the Samples Mine on Cabin Creek [wv]." Photo by Mary Hufford, 1995. Courtesy of the Coal River Folklife Collection, American Folklife Center, Library of Congress.

by rail to power plants, and the resulting wastewater contained coal dust and significant concentrations of heavy metals. The Clean Water Act prevented the release of this slurry into local watersheds, but operators had no efficient way to clean it or dispose of it. As a result, year after year the waste accumulated in massive ponds, often located on steep slopes, held back by earthen dams and artificial liners designed to prevent seepage loss. Across central Appalachia these wastewater lagoons proliferated and grew. Locals throughout the mountains who lived downstream of these impoundments worried that they might become victims of a dam failure, especially after the disaster at Buffalo Creek. As one Martin County resident told a *New York Times* reporter, he was more worried about "what's left up there" in the impoundment than about what had been released in the spill. A complete dam rupture threatened a quick death for downstream residents.[28]

The new forms of Appalachian strip-mining had elevated the consequences of surface mining to new levels. As of 2012, nearly half a million acres of Appalachia were denuded mountaintop or adjacent valleys covered by mining spoils. According to two scholars of mountaintop removal, the technique "has torn apart enormous landscapes across those four states [Kentucky, Tennessee, Virginia, and West Virginia], destroying land and altering the flow of many hundreds of streams."[29] Between the mid-1980s and 2000, approximately seven hundred miles of Appalachian streams were covered in overburden. But for all the obvious consequences of the technique and widespread public opposition, the MTR industry has become a virtually untouchable power in central Appalachia. Coal companies have even applied pressure on public institutions such as the University of Kentucky to open their landholdings to strip-mining. Assorted bills to limit or curtail MTR have recently been voted down by state lawmakers in Kentucky, Tennessee, and West Virginia. Coupled with the growth of "fracking"—using hydraulic pressure to release pockets of natural gas, a process that environmentalists have widely criticized for its potential health and environmental costs—MTR has become "the 'third rail' of Appalachian politics. To touch it means certain political death."[30]

The Martin County spill took place within this complex context of mining history, international energy commerce, politics, and Appalachian land use. But it was undeniably also an unmitigated environmental disaster. Cleanup efforts were complex and long-lasting. The spill eventually covered miles of local watersheds in viscous coal slurry, contaminating the water supplies of a number of towns in addition to harming fish and other wildlife. As mentioned above, in the days following the spill the towns of Fort Gay, Kermit, and Kenova in West Virginia, and Inez and Louisa, Kentucky, all closed their water treatment plants and had water trucked or piped in from outside the spill zone. Communities temporarily shuttered car washes, laundries, and even schools in efforts to conserve water. It was five weeks before all affected town water treatment plants were back on-line.[31]

Efforts to address the spill were varied and costly. Treatment methods included the use of weirs, filters, and small impoundment dams designed to trap the sludge as it moved downstream; vacuum trucks that could skim pollutants from the water's surface; and piping of water from streams, treating it, and pumping it back into the watercourses. Ultimately, cleanup crews had to resort to active, mechanical methods. They systematically dredged the beds of Coldwater and Wolf Creeks to remove settled sludge and excavated the creek banks, hauling away contaminated soil in dump trucks. By December 19, 2000, ten weeks after the spill, cleanup operations had removed almost half a million cubic yards of polluted materials from Coldwater and Wolf Creeks alone. Despite these strenuous efforts, the wave of pollution had by October 19 reached the Ohio River, where the mining waste slowly sank to the bottom of the channel, and its downstream effects became difficult to assess.[32]

The scale of the spill and the subsequent cleanup was enormous, but Environmental Protection Agency (EPA) officials seemed satisfied with the response and MCCC's participation. Despite the fact that mechanical cleanup was ongoing, by mid-December the EPA decided that the daily presence of federal employees was no longer needed on the ground in eastern Kentucky or western West Virginia. Agency officials turned administration of the cleanup over to MCCC, since "conditions no longer represented an emergency as cleanup operations are routine and predictable in nature." At the same time the agency reported that the long-term effects of the spill on local fish populations (and by extension other wildlife and perhaps human populations) remained to be determined by ongoing and future studies.[33]

Not everyone was ready to let the coal company or the EPA off so easily. Local Kentuckians and West Virginians protested that the cleanup was hardly complete. Creeks remained coated in a visible layer of reeking coal sludge, and water supplies were not as clear as they had been before the spill. A documentary film produced by Appalshop, a well-respected regional film studio located in Whitesburg, Kentucky, kept the event in the public eye. *Sludge* (2005), directed by Robert Salyer, connected the Martin County spill to past coal company abuses, such as the Buffalo Creek disaster, and questioned the relationships between coal companies and the politicians who oversaw the EPA. Salyer posed a troubling question: how could a spill that occurred in a populated district in the heart of America, an event thirty times the magnitude of the infamous *Exxon Valdez* spill, so quickly fade from public consciousness?[34]

Grumblings of a cover-up came from within the Martin County investigation too. Jack Spadaro, a member of the Mine Safety and Health Administration (MSHA) investigation of the spill, claimed that his job had been threatened after he declared that MCCC should have anticipated the disaster. The Martin County impoundment had suffered a sizable leak—one hundred million gallons of sludge—just six years before the 2000 breach, and an investigation into that accident had presented MCCC with nine recommendations. The company had largely ignored them, according to Spadaro, and continued to use the impoundment, leading to its inevitable failure. He claimed that the company had certainly known that the impoundment had been leaking and would likely fail at some point in the not too distant future. Of the larger spill, Spadaro noted, "it was just a miracle people didn't die."[35]

The events leading up to the spill were a study in corporate malfeasance, according to Spadaro, but he found the response to the accident even more egregious. He claimed that MSHA officials tried to suppress the investigative report and discredit him when it became apparent that investigators might find MCCC at fault. The federal investigation into company actions ultimately resulted in only two minor charges and a little over $100,000 in fines for MCCC. The penalties were reduced on appeal to just $5,600. After Spadaro complained about the lack of consequences, he was pushed out of his position at the MSHA. The director of the agency who terminated him was a former mining industry employee appointed by Elaine Chao. Chao was secretary of labor at the time and married to Mitch McConnell, a U.S. senator from Kentucky and longtime supporter of the state's coal industry.[36]

The environmental and social disaster that took place in Martin County in 2000 held the potential to open the nation's eyes to the hazards of MTR, but it ultimately made little impact on the public conscience. It did even less to slow the growth of MTR mining. Some county residents and regional activists believed the Martin spill was largely ignored because of where it happened and the identities of those most affected. West Virginia novelist and activist Denise Giardina declared that "if [the spill] had happened somewhere else, everybody would still know about it, and remember it." The relative media silence came because "we're just another country."[37] Spadaro had similar thoughts. He blamed national perceptions of the region for the relative lack of interest, asserting, "It's the *Deliverance* syndrome. People from outside the region still view Appalachia as that dark place over there on the other side of the mountain that you shouldn't even go into. And the people in the Appalachian region are some kind or another backward and deserving of this kind of neglect, and it doesn't matter because they're used to putting up with this."[38] According to this narrative, the idea of Appalachian exceptionalism and its attendant consequences was alive and well in twenty-first-century America.

The spill's county of origin was quite symbolic in this respect. Shortly after announcing a "War on Poverty" in 1964, President Lyndon Baines Johnson had visited Martin County to generate support for his initiatives and to highlight Appalachian poverty as an important national concern. Making his way up a valley about five miles outside of the town of Inez, Johnson met with out-of-work Tom Fletcher and his family on the front porch of their cabin, providing a perfect photo opportunity for the reporters accompanying the president. Here Johnson pledged that the nation had not forgotten the mountain poor. But wealth never came to Martin County, at least not the sort of wealth the president promised, and the experience served to some residents as evidence that their county was more a symbol than a legitimate concern for those in power. Martin County had simply been a convenient place to add white poverty to national worries about the disproportionately black urban poor.[39]

If most of the nation never noticed or quickly forgot Martin County's economic troubles and the sludge spill, many central Appalachian residents believed they had learned a lesson. Although journalist Michael Shnayerson, in an investigation of mountaintop removal, argued that Appalachia's "people are for the most part too poor and too cowed after a century of harsh treatment by King Coal to think they can stop their world from being blasted away," a number of regional residents viewed Martin County as a rallying point.[40] They carried deep disillusionment with the motives of energy companies and with the ability and willingness of county, state, and federal governments to protect their health and interests.[41] One local attributed the root cause of the company's careless actions to its valuation of profits over Appalachian people, lamenting, "It is a shame that it happened. . . . the coal company should have thought more of the people here than to let it happen."[42] Massey did not soothe local dissatisfaction with its claim that the spill was "an act of God," blaming the breach on chance and ignoring the all-too-human origins of strip-mining, coal washing, slurry impoundments, and the old subsurface mining under the site, as well as the warning provided by the earlier spill.[43]

This regional suspicion of the motives of coal companies and government watchdogs found expression in place like Marsh Fork, in Raleigh County, West Virginia. The town's elementary school was located near a Massey Energy coal washing plant and directly below a slurry impoundment holding back almost three billion gallons of coal waste (see fig. 9.5). The impoundment was the same basic type as the one that failed in Martin County, only larger. Parents worried about the cumulative health effects of their children's exposure to air- and waterborne pollutants from coal washing, and the school's water supply was intermittently contaminated. The supreme fear was failure of the impoundment dam, which threatened to re-create the Buffalo Creek disaster. School officials kept a bus parked in front of the school at all times, in case it was needed for an emergency evacuation. Despite growing local protests, Massey officials claimed that the impoundment was safe and that there was no danger to the children from coal washing or waste storage, and government

FIGURE 9.5. "The Sludge Pond at Shumate's Branch, rising above the Marsh Fork Elementary School [*center*], Sundial, wv." Photo by Lyntha Scott Eiler, 1995. Courtesy of the Coal River Folk Life Collection, American Folklife Center, Library of Congress.

agencies failed to intervene. After more than a decade of local protests, including a march from Charleston, West Virginia, to Washington D.C. (a distance of more than 450 miles) by activist and former coal company worker Ed Wiley, concerned citizens raised enough money to move the school to safer ground. Classes began at the relocated Marsh Fork Elementary in 2013.[44]

After the Martin County spill there has also been an increasingly sophisticated campaign of activism attempting to bring national attention to the social and environmental costs of mountaintop removal. Scholars such as Rebecca Scott and Joyce Barry have framed debates over environmental justice and regional strip-mining in modern sociological discourse and gender theories, positing a female nature in the process of being raped and despoiled by masculine, capitalist industry.[45] And activists have found their historians, including

Chad Montrie and Shirley Stewart Burns, who have recorded the evolution of grassroots opposition to strip-mining in Appalachia.[46] A number of other academicians have bridged the university-community gap, through projects bringing scholarly attention to MTR in general and the aftermath of the Martin County episode in particular.[47]

In other forums, nature writer Erik Reece's *Lost Mountain* attempted to raise public awareness of MTR's environmental issues in *Silent Spring* fashion.[48] Novelist Ann Pancake's *Strange as This Weather Has Been* pitted a family with coal-mining roots against the challenges of the new strip-mining landscape.[49] And *Vanity Fair* journalist Michael Shnayerson attacked Massey Energy's record (and its chairman, Don Blankenship) in the best muckraking fashion in *Coal River*, a story about legal battles over strip-mining in southern West Virginia.[50] The Martin County spill and its aftermath also stimulated new activist groups dedicated to halting MTR. The most notable was Mountain Justice, organized in 2004, which brought locals concerned about strip-mining together with such established environmental organizations as the Sierra Club and Earth First! and drew on the energies of college students from inside and outside the region.[51]

Film and the Internet provided even more popular forms of anti–strip-mining activism. Documentary filmmakers created powerful visual imagery and narration challenging the necessity of MTR and the possibility of truly rehabilitating mine sites. In 2005, Appalshop's *Sludge* argued that the disaster response was ineffective, little more than a cover-up, and that the dangers of Appalachian slurry impoundments remained all too real.[52] And in 2011 *The Last Mountain* exposed national audiences to the fight against MTR in southern West Virginia. The film was especially critical of Massey Energy.[53] Online, the website iLoveMountains.org has become an information clearinghouse for people interested in learning more about MTR or searching for ways to become involved in activism. The site even has its own YouTube channel, where some videos have been viewed thousands of times as this is written.[54]

To date, this increased activism has done little to halt mountaintop removal or reduce the threats created by mountain strip-mining, and similar episodes have followed the Martin County disaster. On December 22, 2008, a spill of toxic fly ash slurry of even greater proportions took place at the TVA's Kingston Fossil Plant in Roane County, Tennessee. As in Martin County, a waste impoundment failed, releasing pollutants into the Emory and Clinch Rivers. The total volume of pollutants disgorged into the surrounding environment was approximately 1.1 billion gallons, creating a spill more than three times the magnitude of the Martin County event and the largest in the history of the United States.[55] Six years later, in 2014, a similar spill took place at the Dan

River Steam Station, a Duke Energy power facility located near Eden, North Carolina. This breach released just under forty thousand tons of coal ash and approximately twenty-seven million gallons of contaminated water into the Dan River, which drains and provides drinking water to communities along the Virginia and North Carolina border. By some estimates the Dan River incident was the third-largest coal waste spill in U.S. history, with economic damages approaching $300 million. Although the plant was located in the Piedmont foothills east of the Blue Ridge, like most other southeastern power plants it heavily relied on strip-mined Appalachian coal for fuel.[56]

ↄ

What did the Martin County spill reveal about Appalachian environments at the turn of the twenty-first century? First, it made clear that old challenges to Appalachian land use and conservation remained concerns. Coal companies like Massey Energy were still powerful forces, perhaps central Appalachia's most important private economic players. Natural resource extraction had taken new forms in the mountains: mountaintop removal had increasingly replaced underground mining, waste storage facilities covered large tracts of land and posed new pollution and regulatory challenges, and the clean coal campaigns attempted to change public perceptions of the traditional fossil fuel. But underneath the new techniques and sophisticated language were relationships between economic power and nature that extended back more than a century. Coal, extracted at the intersection of local miners' labor, capital investment, and mountain environments, continued to move in an endless stream along Appalachian railways, largely to fuel the never-ending energy needs of the outside world.

Denise Giardina, Jack Spadaro, and similar public voices see these challenges as uniquely Appalachian. Americans made uncomfortable by (or perhaps feeling guilty for) the "backwardness" of the southern mountains have looked away from the costs and hazards of mountaintop strip-mining. Doubling down on this extractive bonanza, energy companies have also ramped up natural gas drilling in the Marcellus shale belt that underlies much of West Virginia, eastern Ohio, and western Pennsylvania, and people outside the mountains have surrendered the long-troubled region as a "national sacrifice zone."[57] Perhaps. In the face of the region's history this is a satisfying account. But the idea of an Appalachian sacrifice zone as an outgrowth of regional history is in some ways too neat. After all, most Americans close their eyes to a wide range of environmental consequences stemming from their modern lives. They pay relatively little attention to the results of hydraulic fracturing in Texas, the Dakotas, and Utah, the pit mining wastelands covering the

Alberta tar sands, and the hypoxic zone spreading from the mouth of the Mississippi River to deaden a growing swath of the Gulf of Mexico. Much of rural America and indeed the rest of the rural world are sacrifice zones of some sort. In Americans' blindness to the consequences, Appalachian communities dealing with the effects of MTR are perhaps more typical than they are exceptional.

The Martin County disaster also highlighted new, more sophisticated ways people had learned to fight back in Appalachia. To be sure, criticism of the spill, and of mountaintop removal more broadly, contained conventional elements of reactionary NIMBYism ("not in my backyard") as well as the personal outrage and angry frustration that contributed to bloodshed in early twentieth-century Appalachian coal protests. But MTR opponents also employed new language and sophisticated organizational skills forged in the long decades of regional poverty and labor activism. Activists developed important relationships with national organizations not traditionally associated with Appalachian causes, from the Sierra Club to Earth First! Appalachian people also learned to speak the language of middle-class environmental aesthetics. They appealed to desires for beauty, health, and permanence (in the language of one historian of modern environmentalism).[58] They also used modern communication methods as effective weapons, using the Internet in particular to spread Appalachian messages to a global audience.[59] These tactics made visible the ties that had always existed between the people and natural resources of the southern mountains and the world beyond. They highlighted the sometimes-problematic ways in which nature connected rather than isolated Appalachia. In protesting the Martin County spill, local people concerned about the Appalachian environment's present and future successfully accomplished what they had tried so long to do: they defined their place in the world.

෨

Blair Mountain in Logan County, West Virginia, was the site of one of Appalachia's and the nation's most seminal labor clashes. In 1921, as many as seventy-five hundred coal miners assembled in the southern mining districts to march on the state capitol in Charleston to protest company abuses and assert their right to unionize. Guards employed by the companies and local law enforcement officers, also funded largely by the industry, met the armed miners at Blair Mountain, and a battle ensued. The arrival of federal troops in support of the companies and White House assertions that continued violence was tantamount to treason ended the shootout. The companies won at Blair Mountain, at the cost of dozens of lives, but the conflict brought national attention to the violence and corruption of the central Appalachian coalfields and would eventually contribute to hard-won labor reforms.[60]

Today another sort of battle rages over Blair Mountain. On one side are companies that own land and mineral rights on the site, including subsidiaries of Arch Coal and Alpha Natural Resources (the new name of a reorganized Massey Energy), which want to remove the mountain's top to get at the rich coal seams underneath it. They argue that the techniques of mountaintop removal are the only way to efficiently mine Blair Mountain, and they argue for the sanctity of the right to use private land to turn a profit. On the other side is an array of public interest groups that assert that the site is far more valuable for its environmental and historic qualities than its coal. They argue that Blair Mountain is a public treasure that must be saved from destruction. Opponents of strip-mining pushed to have the site listed on the National Register of Historic Places—it was listed and then quickly delisted because of landowner opposition—and have filed suit to prevent the companies from initiating mining. In 2014 a federal appeals court ruled that the various groups did have the right to initiate legal actions, and as of this writing the future of Blair Mountain remains in the courts' hands.[61]

What happens at Blair Mountain may be a harbinger of the future of both mountains and mining in Appalachia. Mountaintop removal has largely replaced traditional underground mines in the region, today producing more coal than the older methods and promising to increase its share as the most easily accessible coal seams disappear while demand for energy grows. A national determination to rely less on foreign oil also promises to encourage power production fueled by the "clean coal" of places like Logan and Martin Counties. The techniques of MTR have already transformed the regional environment and economy, critics would say for the worse, and at places like Blair Mountain MTR threatens to physically erase Appalachia's past as well. The ground that miners and operators fought over might literally disappear in the dust of explosions and under the blades of bulldozers, in a second victory for the coal companies at Blair Mountain. If it does, the benefit will be relatively cheap electricity; the losses await calculation.

Epilogue

The Adelgid and the Salamander

APPALACHIA'S ENTANGLEMENT with distant environments contin-
ues today, increasing each year. As always, these ties extend well beyond the
human realm and into networks of biological interaction. One illustration of
this is the region's ongoing struggles with a long legacy of biological exchange
between different areas. These long-range transfers—often referred to as the
Columbian Exchange by historians—began at Columbus's landing in His-
paniola, continued through the work of settlers and agents of empire (like
the botanical collectors who roamed Appalachia two centuries ago), and were
perpetuated time and again thereafter by the economic exchanges and travel
that accelerated with the Industrial Revolution.[1] Most often associated with
exchanges between continents, the turmoil of the Columbian Exchange and
its global action also accelerated the volume and pace of regional exchanges
and commoditization, in Appalachia introducing species from the lowlands
as well as from across the ocean. Today Appalachia is still home to an astonishing array of specialized plants and animals, each evolved to exploit an
ecological niche in the region's exceptionally varied landscapes. Some of them
are quite rare, defined as "endangered" in modern parlance. The movement
of people and trade in commodities has introduced many new species too,
some of which have flourished, and in the process challenged older ecological
relationships to such an extent that people have come to call them "invasive."
Exploring the ongoing ties of endangered and invasive species is a fruitful way
to witness a modern manifestation of the region's long history of connection.
One enlightening example of this intersection takes us again to Grandfather
Mountain, where a rare salamander and an increasingly common Asian insect
now occupy the same ground on a site often considered remote but that has a
deep history of contact with distant places.

The wilder corners of the Appalachian environment are perhaps best known
for their impressive forests and "charismatic megafauna," such as bears, deer,
and mountain lions (which may or may not be locally extirpated), but they also

contain a tremendous diversity of amphibians within their bounds, no matter how they are drawn.[2] By some estimates, a typical plot of Appalachian forest contains more salamanders by weight than all other animal species combined. Grandfather Mountain, with its significant elevation range and habitat variation—from shady, wet coves to arid and windswept granite outcrops—exemplifies this amphibian abundance. The modern state park boundaries encompass a number of endangered, threatened, or unique salamanders (among the twenty-one confirmed salamander species currently inhabiting the mountain). Rare species include Yonahlossee salamanders, Blue Ridge two-lined salamanders, hellbenders (the largest of all North American species), pygmy salamanders, and Weller's salamanders. Ongoing scientific studies within the park contribute to modern understandings of these species' natural histories.[3]

Salamanders, like all creatures whose lives touch those of humans, also have cultural histories. Grandfather's varied habitats and biota made it one node in the empire of scientific collecting that spanned the early modern era, as men of science (they were almost always men) journeyed the globe naming, collecting, and selling species, along the way accruing power and wealth for themselves and the European imperial powers that supported them.[4] As noted, the mountain had been a popular site to visit for botanizers like John Lyon and geologists like Elisha Mitchell.[5] A century on, as the biological sciences specialized, the mountain's abundant amphibians attracted pioneering herpetologists, still imagining fame and fortune from the discovery of species hitherto unknown to western taxonomy texts. Most notable among these herpetologists (and certainly the one with the most dramatic story) was the wunderkind Worth Hamilton Weller, a Cincinnati native who had made a mark on the scientific community by 1930, only midway through his high school career. In the summer between his sophomore and junior years, Weller discovered, described, and named two salamander species new to science. The first of his finds, *Gyrinophilus porphyriticus duryi*, was a Kentucky cave dweller. He found the other, the eponymous Weller's salamander (*Plethodon welleri*), in damp soil under the stones littering the upper reaches of Grandfather Mountain.

Weller returned to the mountain on another collecting trip within a week of his high school graduation, when his story turned from precocious to tragic. While collecting specimens on a stormy day he apparently slipped and fell to his death from one of Grandfather's rugged cliffs. A search party found his body—next to a sack containing several of his namesake salamanders—four days later, pinned between two boulders.[6] Despite his intriguing story, Weller has been largely forgotten, but his salamander lives on, one of the region's most endangered species. Today a hiker with a field guide and some patience can re-create Weller's moment of discovery by finding the small amphibians, notable for their silver herringbone-patterned backs, under rocks high along Grandfather's ridge line. Searches nearby might also produce other endangered salamanders. The mountain's wet draws could reveal minuscule pygmy salamanders in search of their invertebrate prey, while the leaf litter downslope is home to larger Yonahlossees, impressive, active salamanders with an erect posture and mottled red backs. A descent all the way downslope to the headwaters of the Watauga River or Linville River might uncover a two-foot-long hellbender under a flat rock in the swift-flowing water.

Like so much animal and plant life in the increasingly popular Blue Ridge Mountains, now more a landscape of vacation homes than the backwoods domain of isolated Scots-Irish settlers mythologized by early regional scholars, many salamanders on Grandfather and similar peaks face pressure in the form of habitat change. Much of this change is the result of development, in Grandfather's case roads, two golf courses, and second homes on its lower flanks, and a tourist attraction located on one end of its central ridge. Grandfather and similar areas have become landscapes of commoditized recreation, where businesses package and sell outdoor lifestyles and activities to tourists and retirees. But perhaps just as important have been a number of invasive species that have accompanied regional human endeavors. Pygmies, Yonahlossees, and Weller's salamanders have repeatedly had to adapt to new species moving into their habitats and changing local ecologies, altering the spaces where multiple species live.

The region's most famous invasive (besides humans) was the chestnut blight, *Cryphonectria parasitica*, a fungus of Asian origin. Trade in nursery stock accidentally carried the blight to North America, where foresters first documented the parasite in New York in 1904, and it quickly spread southward. Within half a century it had killed every mature chestnut tree in the eastern United States, perhaps three and half billion individuals, leaving their ghostly trunks to fall and slowly rot over the following decades. This ecological disaster radically changed the composition and function of Appalachian forests, and it altered the livelihoods of people who depended on chestnut mast for feeding livestock and for sale to urban markets.[7] Never before in Appalachia had entire ecosystems experienced such a rapid transformation. Historian Kathryn Newfont has argued that the "importance of the chestnut loss is impossible to calculate fully and all but impossible even to fathom."[8] Sociologist Donald Davis summarizes the significance of chestnut die-off most succinctly in his description of the event as "a whole world dying."[9] The fungus touched all corners of the Appalachian world as chestnut canopies disappeared, leaving voids slowly replaced by other plant species. Salamanders, denizens of the forest floor and reliant on shade and soil moisture, must have been affected by these ecological transitions, but like so much about their past, exactly how they struggled with the changes remains hidden. Yet we know they persevered.

For every species fast disappearing from the Appalachian Mountains, a new one (or two or three) moved in to take its place. Other motile species followed the chestnut blight, though they have yet to secure a place in the regional imagination or historiography to the same degree as the blight. In the highest reaches of Appalachia, a few decades after the chestnuts withered and died, an invasive insect transformed the forest once again, altering one of the few habitats left untouched by the earlier fungal blight. This time the invasive species was a primitive form of European aphid, the balsam woolly adelgid (*Adelges piceae*), which first drew notice in Appalachia on Mount Mitchell, North Carolina, in 1957, though it had likely existed within local forests for twenty years or more by that time.[10] The balsam woolly adelgid (BWA) feeds on fir needles with grave results for the tree. As one historian explains its habits and the results: "Its saliva stimulate abnormal growth, causing fir stems to swell and twist at odd angles. Such changes slow the movement of water and minerals through the tree's heartwood so that an infested fir, in essence, starves to death, usually in three to nine years."[11] Mount Mitchell's trees rapidly succumbed and died, and BWA was on the march, floating on wind currents and hitchhiking on songbirds in search of new host forests. By 1963 the adelgid had found Grandfather Mountain's fir stands.[12]

The BWA infestation had consequences that extended beyond its effects on tree populations. As was the case with chestnut die-off, fir mortality triggered significant ecological changes on many southern Appalachian mountaintops. When firs perished and fell, neighboring red spruce suffered greater wind damage. The resulting thinner canopy cover challenged understory species that had relied on the shady protection of taller vegetation, the forest floor became warmer and drier, and the result was less attractive habitat for various insect species that anchored local food chains.[13] The consequence, as one ecologist writes, was (and remains) a forest "in a state of structural and compositional reorganization."[14] That fir death changed salamander habitats is certain, but to exactly what extent remains less than concrete. Other environmental changes taking place concurrent to the spread of BWA complicate the ecological calculus. What climate change, ongoing natural shifts in forest composition, acidic deposition from pollution-laced clouds, and other invasive species meant and will mean for amphibian populations often amounts to educated guesswork (and a lot of repetitious surveying). Despite these variables, for salamanders that rely on moist forest floor environments—such as Weller's and pygmies— the removal of firs from high-elevation woodlands no doubt posed serious challenges. Again, though, our knowledge concerning how Weller's salamanders adapted to these changes remains incomplete. For example, while reduced canopy cover must have proven threatening, populations might have temporarily benefited from an abundance of dead fir trees, as females prefer rotting conifers as nest sites.[15] All that can be said with certainty is that the species survived the forest transitions that accompanied the success of BWA.

BWA was never truly contained. Identification of the insect was tardy, the adelgid reproduced too rapidly and spread too easily, and the human effort to control it was organized too slowly. State park officials at Mount Mitchell aggressively attacked the outbreak with the powerful pesticide lindane and then introduced predatory insects in an effort to check the spread of BWA, but it was all to little effect, as nearby untreated stands of firs served as continual reservoirs of reinfection.[16] During the 1980s the adelgid spread throughout the Great Smoky Mountains National Park, killing 90 percent of the park's mature Fraser firs, and by the twenty-first century virtually every fir stand in Appalachia harbored BWA.[17] The insect seems to have become a permanent Appalachian resident, one almost as effective at its work as the chestnut blight.

But the march of mobile species across Appalachia and through Grandfather's forests did not halt with the BWA. Among the region's late twentieth-century invaders was another tiny aphid, an evolutionary relative of the BWA, the hemlock woolly adelgid (*Adelges tsugae*). In a situation bearing eerie similarities to the establishment and spread of BWA, the hemlock woolly adelgid (HWA) first appeared in the vicinity of Richmond, Virginia, in 1951, again as an accidental import caught up in international commerce, in this case on East Asian plant stock.[18] HWA soon spread across and then beyond the boundaries of the Piedmont South, moving into the hemlock groves of Appalachia, habitats replete with its preferred food source. HWA attacked both Appalachian hemlock species (the eastern hemlock and the less common Carolina hemlock), drying out and killing trees in a manner similar to BWA. In the region "all hemlock stands appear[ed] to be equally susceptible to HWA attack."[19] The spread of HWA accelerated in the 1990s and first decade of the new millennium. By 2008 roughly one-third of all hemlocks in southern Appalachia were dead, and current mortality approaches 100 percent in infested areas.[20] HWA has proven as efficiently deadly to its food sources as the BWA infestation that preceded it, and both adelgid species had proven themselves powerful agents of change.

Here I must admit to a lack of objectivity or at least to a personal interest in this particular history. I was temporarily a part of this story, and my experience with HWA was hands-on. I worked as a natural resource manager at Grandfather Mountain from 2001 until 2004, a period that coincided with HWA's arrival in the area. With the experience of the ecological disturbances resulting from BWA influencing management decisions, Grandfather's staff faced the new invasive with the determination to save as many hemlocks as possible. The park contained significant stands of mature trees as well as a few old-growth groves in its more remote reaches. After a good deal of research, we settled on a cutting-edge treatment to combat the invasive insect.

Our plan, impossible as it sounds, was to inject every hemlock tree larger than a sapling with a pesticide that would hold the adelgid at bay for two years, at which time the trees would need reinjection if adelgids remained active in the area. The chemical we used was imidacloprid, a nicotine-like substance commonly applied on agricultural fields and used to keep fleas and ticks off domestic animals. (Readers who have used a tube or capsule of flea and tick repellent on their cats or dogs have likely applied imidacloprid.) We considered other options, ranging from biological controls (such as introducing predator species) to applying harsh bifenthrin pesticide sprays, but we determined each to be impractical, cost-prohibitive, or too environmentally toxic. Even the targeted use of imidacloprid was a desperate and costly attempt to preserve a portion of the hemlock population. The ultimate hope was that the adelgid irruption would burn itself out while the park's hemlocks were protected, at which point imidacloprid treatment could cease. In other words, we hoped adjoining environments would change faster than our own, as neighbors with fewer resources were unable to inject and then would lose their hemlock stands. Grandfather might be protected if that happened, with one source of its connectivity to the broader world broken by a moat of dead hemlocks.

The logistics of the endeavor were daunting. We had to locate each hemlock in the park, often by using aerial photographs, hike to the location, measure the tree's diameter, and then calculate the appropriate chemical dosage. Next we used cordless drills to breach the tree's thick, plate-like bark and penetrate the phloem, drove plastic tubes into the holes, and attached pesticide capsules to each tube. The hemlock's natural vascular action carried the imidacloprid up the tree and into each needle, where feeding adelgids would absorb the insecticide, which then inhibited their ability to eat and reproduce. Next we tagged the tree and recorded the date and dosage and tag number in a field log. A few days later, once the tree had absorbed all of the pesticide, we removed the empty capsules and tubes and plugged the drill holes in an effort to reduce the risk of bore-site infections. We followed these steps thousands of times in some of the roughest terrain in the southern mountains.

Surprisingly, a few stands of old-growth hemlock on Grandfather had survived the onslaughts of historic logging operations, forest fires, and park development, sheltered at the upper reaches of impossibly steep hollows or growing on boulder fields where the trunks squeezed between stones as large as New England cottages. It was in these remnants of the historic Appalachian cove forest where our work was most critical and most difficult. We scrambled through laurel hells and over razorback ridges to treat each of these stands, lugging drills, sacks of pesticide capsules, tape measures, logbooks, lunches, and water bottles. And then we repeated these journeys a few days later, this time hauling bags of empty capsules and tubes. Some spots, especially the tangles of mossy great rhododendron, were almost impossibly thick. To get through such a thicket, fisherman and adventurer Philip Kennedy wrote more than a century and a half earlier when tracking through similar forests in West Virginia, required "such pulling and tugging—such twisting, plunging, breaking, crashing, and tearing—'I never remember ever to have heard'—or seen. . . . To tell how at last we all did get out, overtaxes any powers of description that I possess."[21] Not much has changed in this regard. Over the course of two summers we managed to inject nearly every tree, and the mountain's hemlocks have since received additional treatments, but the jury remains out on whether or not these efforts will ensure the survival of the species on Grandfather. All the tugging and twisting may have been for naught. The adelgid continues to feed on Appalachian hemlocks, slowly drying up the forest sentinels where they stand.

Hemlock survival is of special importance in Appalachian forest ecosystems because Grandfather and similar sites are not in stasis. They are still transforming because of the insect and fungal infestations that preceded the spread of HWA, brought on by Appalachia's environmental connectivity. The actions of BWA had changed higher elevation habitats, perhaps pushing some species to the margins of their ranges where hemlocks provided new shelter, and the earlier chestnut blight had all but eliminated the region's most numerous tree species, likely resulting in the gradual movement of hemlocks into ecological niches once occupied by chestnuts. Hemlock forests had served as a refuge between the damaged spruce-fir forest above and devastated chestnut woods downslope, but the refuge is proving temporary.[22]

Hemlock death changes the forest canopy in much the same manner (at least in general terms) as did widespread Fraser fir death. John Quimby, an entomologist, explains: "There are no darker, cooler places in the forest than under a hemlock canopy."[23] As hemlock needles yellow and fall, followed eventually by rotten limbs and trunks, more light reaches the forest floor, with consequences for ground-level nutrient concentrations, stream temperatures, and soil moisture. And without a thick canopy to slow raindrops and without hemlock roots to hold waterlogged soil in place, erosion and flooding become increasing concerns throughout Appalachia (as well as farther downstream in the Piedmont and coastal plains). Such changes are of obvious importance not only to people living in these watersheds but also to other species that rely on mature hemlock forests for survival.[24]

One such species may be Weller's salamander. Although the amphibian was first described in high-elevation spruce-fir forests and is often associated with them, subsequent studies found the salamanders at lower elevations, down to twenty-five hundred feet in Tennessee, well into the hemlock and hardwood forest.[25] As at higher elevations, in these sites the salamanders use the moist forest floor for cover and rely on rotting conifers for nesting sites, so dense canopy cover and a continued growth of hemlocks seem of great potential importance here.[26]

Another amphibian at risk as the hemlock forest succumbs to HWA is the Yonahlossee salamander. The Yonahlossee, named for the road carved along the eastern slope of Grandfather Mountain to connect Linville Resort to Blowing Rock at the end of the nineteenth century, can today be found on moderate- to high-elevation slopes (above fourteen hundred feet) from western North Carolina to southwestern Virginia. Like Weller's salamanders, Yonahlossees depend on moist forest floors, and their chief threat is habitat alteration, whether in the form of forest clearance for residential development or canopy die-off.[27] The HWA has the potential to compound the critical habitat changes wrought by BWA, leaving vast stretches of Appalachian coniferous forest forever altered and potentially pushing Weller's and Yonahlossee salamanders to the brink of extinction. This situation is partly a biological struggle: adelgids seek resources for feeding and reproduction, while salamanders pursue their lives in the leaf litter, where shade and moisture are critical. In other ways it is cultural: people struggle to maintain aesthetic values and keep a landscape of impressive old-growth hemlocks and rare amphibians in stasis. By either definition, it is a product of a history of environmental connections tying Grandfather to distant places.

The stories of various salamander and adelgid species, even at sites such as Grandfather where they intersect, might seem disparate events, but they are connected in ways worth considering through their common engagement with the broader environment, an engagement shared with humans. The mountain, like all other corners of Appalachia, has become a meeting ground where a wide variety of living creatures intersect. People have worked in undeniable ways to connect this space to more distant ones: they moved nursery stock across oceans, built homes and resorts, and came to take in the scenery and fresh air. As a part of all these actions, they purposefully as well as inadvertently transported insect, fungal, bacterial, and viral passengers. And humans worked to limit the spread of insects they deemed undesirable, to count salamanders, and to "improve" habitat for species they designated worthy of preservation. Additional connections between Appalachia and other regions continue to shape the mountain environment. Power plants in the Ohio and Tennessee River valleys (fired with West Virginia and Kentucky coal) emit pollutants that alter regional climate and rainfall, ongoing road-building accelerates runoff erosion and introduces more tourist traffic, and federal money simultaneously funds various area development and conservation programs. The case of adelgids and salamanders on Grandfather also reveals how these interactions often produced unforeseen results. Adelgids were astonishingly successful at achieving their ultimate biological imperative: reproduction. For all the seeming power of humankind—their institutions and chemicals and machinery—people could do little to slow the insects. Salamanders' actions are perhaps subtler yet ultimately may be the most impressive. For all the demonstrable power of people and adelgids in this particular place, Weller's, Yonahlossees, and other imperiled salamanders have found a way to adapt and survive, at least so far. This is no mean feat.

The defining feature of this story is the persistent if ever-changing nature of Appalachian connections. Like ecosystems, these relationships have rarely had neatly fixed physical or temporal boundaries. The movement of the chestnut blight, BWA, and HWA ran one into another and shifted across fluid landscapes, becoming entangled with the mountains' histories of farming, forestry, tourism, and recreation. Other invasive species are busily working to reshape Appalachia too, colonizing the mountains as effectively as the rest of a globe now entangled in international commerce. Some of them may yet become important players in Grandfather's future. These invasives include Dutch elm disease, gypsy moths, dogwood anthracnose, and emerald ash borers, which are all acting on their own behalf and shaping the land in the process.

Just when the story seems set, nature acts in surprising ways, leaving this narrative open-ended. For all the ecological tumult in fir and hemlock forests, Weller's and Yonahlossee salamanders remain part of these environments. They have found ways to survive grave challenges, and they have done so largely without human assistance. Plant communities too have seemingly adapted in response to adelgid infestations. Vigorous stands of young Fraser firs blanket spaces where mature trees succumbed, although it remains unclear if this regrowth will reach maturity and, if it does, how long it can resist BWA. A few botanists believe Fraser fir populations are beginning a true recovery.[28] In the hemlock forest, meanwhile, dying trees are in the process of yielding territory to another shade-providing species, great rhododendron (*Rhododendron maximum*). Just as hemlocks moved into ecological niches vacated by dying chestnut trees in the early and mid-twentieth century, rhododendron seems to be colonizing gaps left by infested hemlocks, shading the forest floor and perhaps shielding its denizens, at least until the next invasive species arrives.[29]

&

What does this particular ongoing struggle to balance human needs and the resulting environmental changes tell us about the legacy of Appalachia's past? It demonstrates that the connections and commodity flows so central to regional history continue to remain important today, perhaps more so than ever before. It is a reminder of the mountains' long, continuous interaction with the broader world, despite ongoing stereotypes of regional isolation. For Weller's and other threatened high-elevation salamanders, the movement of regional air masses is of critical importance, and no physical curtain divides their Appalachian habitats from the sources of industrial pollution dotting the Ohio and Tennessee Valleys. Every shower that falls on the top of Grandfather—which, thanks to air pollution, has a pH of Mountain Dew and includes a stiff dose of aluminum—is a conversation between flatlands and salamanders. It is Appalachian coal that has and continues to fire many of those electric plants, and in some cases Appalachian people who left the mountains to work them or to work in the factories they powered. And coal power plants in those regions in turn link to consumers and economic chains across the continent. The arrival of the adelgid to the mountain's slopes is part of even more far-flung exchanges, the global traffic in plant species and their accompanying pests. The HWA is merely today's version of earlier biotic revolutions that shook the mountains, driven by gypsy moths, the balsam wooly adelgid, and, decades before them, the chestnut blight and the smallpox virus that forever transformed Appalachian cultures and forests.

There is a temptation to be pessimistic about the fate of Appalachian salamanders and about the regional environment's history and its future. Even in this age of environmental awareness, these exchanges remain full of inadvertent consequences. Who could have foreseen that a complex mix of scenic tourism, logging, and the business of horticulture would deliver an insect pest and threaten a rare salamander? And connectivity can highlight ongoing inequalities of power. The efforts to combat HWA with chemical treatments is one in a long line of attempts to manage nature that relied on disparate wealth and resources to alter the environment. The use of imidacloprid on Grandfather stands a chance in part because not every landowner can afford to use this method. Grandfather's trees have merely to survive long enough for all the surrounding untreated hemlocks to die and thus deplete the food source for the adelgid. Although the efforts to preserve a key species of the mountain forest in a least a few locations might be considered noble, they highlight the ongoing disparity embedded in conservation and environmental management in Appalachia. Large companies, federal and state governments, and wealthy outsiders continue to wield disproportionate power when it comes to the manipulation of nature. Massey Coal can strip mountaintops and exploit labor, parks can combat invasive species with methods unfeasible for independent landowners, and vacationers can purchase gated properties that monopolize viewsheds and enclose historic commons. The commoditization of Appalachian nature continues. Some of these inequalities work against environmental stewardship, and others appear to aid conservation, but all threaten to destabilize the social relationships that are critical to the future of the mountain environment.

Neither the HWA nor Weller's salamanders are commodities, per se, but their lives have become entangled with several of the flows this book examines. Their current state is a product of an international commerce in plant specimens, with roots in the period of botanical empire that sent men like John Lyon and André Michaux into the Appalachian Mountains. The forests that both inhabit were shaped by the commercial logging that began in the late nineteenth century. The adelgid spread with such rapidity thanks to the roads and rails that helped build regional centers like Roanoke and Oak Ridge. Human efforts to combat the adelgid have derived much more motivation from a desire to maintain profitable recreational landscapes such as Grandfather than from concern about a rare species of salamander. And the struggle to contain the invasive insect has played out in part on federal and state lands and has involved a trickle of the flood of government dollars that flows through the region.

These ties between salamander and adelgid illustrate the core message of this book. Appalachia has long been as connected as it has been isolated, and the forces blamed for its isolation—the region's rugged geography and environments, shaped over eons by the geologic forces of mountain uplift followed by water and wind erosion—also encouraged those connections. The earth's processes formed thick forests full of deer and varied plant life. They deposited gold, salt, coal, and iron ore. And they produced fertile valleys, swift-flowing streams, and ridges with beautiful views overlooking wave after wave of blue-green mountains. Even as the resulting topography made travel and communication difficult, these natural resources attracted people intent on using them to human advantage.

On Grandfather you can stand in a spot and be at one center of these connections. Find an old-growth hemlock, twenty feet around and two hundred feet tall, its plate-like bark a reddish-brown in the spots where moss does not grow. Look up, and you can search for the cottony residue that signals an adelgid infestation and a perhaps irrevocable alteration of the future forest.

Look down, turn over a few rocks, and you might uncover one of a number of salamander species. The invasion of one directly affects the prospects of the other, and their lives are entwined in the site's human history, as much a product of commerce in ideas and commodities as they are reliant on ecological cycles. Today, as has been the case for the past four hundred years, even as these exchanges highlight the diversity of Appalachia, they also serve to knit it together. How these exchanges will proceed remains to be seen.

NOTES

Introduction. A Constant Arcadia

1. For a good introduction to the region's varied environments (although one focused on the Blue Ridge portion of Appalachia), see Steve Nash, *Blue Ridge 2020: An Owner's Manual* (Chapel Hill: University of North Carolina Press, 1999). The botanical bible for southern Appalachia is Albert E. Radford, Harry E. Ahles, and C. Ritchie Bell, *Manual of the Vascular Flora of the Carolinas* (Chapel Hill: University of North Carolina Press, 1968).

2. For the argument that we need no longer critique the idea of Appalachian exceptionalism, see Steven E. Nash, *Reconstruction's Ragged Edge: The Politics of Postwar Life in the Southern Mountains* (Chapel Hill: University of North Carolina Press, 2016), 5–6.

3. James Watt Raine, *The Land of Saddle-Bags: A Study of the Mountain People of Appalachia* (1924; repr., Lexington: University Press of Kentucky, 1997), 19.

4. Richard Follett, Sven Beckert, Peter Coclanis, and Barbara Hahn, *Plantation Kingdoms: The American South and Its Global Commodities* (Baltimore: Johns Hopkins University Press, 2016), 12.

5. Wendell Berry, *The Unsettling of America: Culture and Agriculture* (San Francisco: Sierra Club Books, 1996), 6.

6. Good places to begin a reading of existing regional coal scholarship include Alessandro Portelli, *They Say in Harlan County: An Oral History* (New York: Oxford University Press, 2010); Rebecca J. Bailey, *Matewan before the Massacre: Politics, Coal, and the Roots of Conflict in a West Virginia Mining Community* (Morgantown: West Virginia University Press, 2008); Crandall A. Shifflett, *Coal Towns: Life, Work, and Culture in Company Towns of Southern Appalachia, 1880–1960* (Knoxville: University of Tennessee Press, 1991); John Gaventa, *Power and Powerlessness: Quiescence and Rebellion in an Appalachian Valley* (Urbana: University of Illinois Press, 1982); Ronald D. Eller, *Miners, Millhands, and Mountaineers: Industrialization of the Appalachian South, 1880–1930* (Knoxville: University of Tennessee Press, 1982); and Harry M. Caudill, *Night Comes to the Cumberlands: A Biography of a Depressed Area* (Boston: Little, Brown, 1962).

7. Kate Brown, *Dispatches from Dystopia: Histories of Places Not Yet Forgotten* (Chicago: University of Chicago Press, 2015), 6.

8. Almost every study of the region begins with a seemingly obligatory justification of its limits. For an expansive definition of Appalachia, one that extends from western New York to northwestern Mississippi, see the Appalachian Regional Commission's official map of Appalachian counties, www.arc.gov/images/appregion/AppalachianRegionCountiesMap .pdf. For an insightful discussion of the region's boundaries, see John Alexander Williams, *Appalachia: A History* (Chapel Hill: University of North Carolina, 2001), 8–14. This book focuses on the mountains from northern West Virginia to northeastern Alabama, but, as a study of environmental connections, ideas, and the flows of goods, these boundaries are necessarily blurry.

9. Donald E. Davis, *Where There Are Mountains: An Environmental History of the Southern Appalachians* (Athens: University of Georgia Press, 2000).

10. On the etymology of the name of the region, see Williams, *Appalachia*, 19–20.

11. For a start in world-system theory, see Immanuel Wallerstein, *The Modern World-System: Capitalist Agriculture and the Origins of the European World-Economy in the Sixteenth Century* (Waltham, Mass.: Academic Press, 1976); Andre Gunder Frank, *The World System: Five Hundred Years or Five Thousand?* (London: Routledge, 1994); and Immanuel Wallerstein, *World-Systems Analysis: An Introduction* (Durham: Duke University Press, 2004).

12. John Alexander Williams, "Appalachia as Colony and as Periphery: A Review Essay," *Appalachian Journal* 6, no. 2 (Winter 1979): 158.

13. An important inspiration for this take is David C. Hsiung, *Two Worlds in the Tennessee Mountains: Exploring the Origins of Appalachian Stereotypes* (Lexington: University Press of Kentucky, 1997).

14. William Cronon, "A Place for Stories: Nature, History, and Narrative," *Journal of American History* 78, no. 4 (March 1992): 1347–76.

15. Henry Shapiro, *Appalachia on Our Mind: The Southern Mountains and Mountaineers in the American Consciousness, 1870–1920* (Chapel Hill: University of North Carolina Press, 1978); Allen Batteau, *The Invention of Appalachia* (Tucson: University of Arizona Press, 1990).

16. The famous phrasing here is William Goodell Frost's, first presented in "Our Contemporary Ancestors in the Southern Mountains," *Atlantic Monthly* 83, no. 3 (March 1899): 311–19.

17. One of the best summaries of early stereotyping remains Cratis D. Williams, "The Southern Mountaineer in Fact and Fiction" (PhD diss., New York University, 1961), reprinted in installments in the 1975 and 1976 issues of *Appalachian Journal.*

18. *Buckwild* (Parallel Entertainment Pictures, 2013); *The Wild and Wonderful Whites of West Virginia*, dir. Julien Nitzberg (Dickhouse Productions, 2010).

19. *Wrong Turn*, dir. Rob Schmidt (Summit Entertainment, 2003).

20. Jeff Biggers, *The United States of Appalachia: How Southern Mountaineers Brought Independence, Culture, and Enlightenment to America* (New York: Shoemaker & Hoard, 2006).

21. J. D. Vance, *Hillbilly Elegy: A Memoir of a Family and Culture in Crisis* (New York: Harper, 2016), 3, 4.

Chapter 1. Leather: The Deerskin Trade in the Southern Mountains

1. Henry Timberlake, *The Memoirs of Lieut. Henry Timberlake* (London: J. Ridley, W. Nicoll, and C. Henderson, 1765), 30.

2. Larry R. Kimball, "Early Archaic Settlement and Technology: Lessons from Tellico," in *The Paleoindian and Early Archaic Southeast*, ed. David G. Anderson and Kenneth E. Sassaman (Tuscaloosa: University of Alabama Press, 1996), 159.

3. Good places to start in this literature include Kathryn E. Holland Braund, *Deerskins and Duffels: The Creek Indian Trade with Anglo-America, 1685–1815*, 2nd ed. (Lincoln: University of Nebraska Press, 2008); Heather A. Lapham, *Hunting for Hides: Deerskins, Status, and Cultural Change in the Protohistoric Appalachians* (Tuscaloosa: University of Alabama Press, 2005); John Oliphant, *Peace and War on the Anglo-Cherokee Frontier, 1756–*

63 (Houndsmill, UK: Palsgrave, 2001); Shepard Krech III, "Deer," chap. 6 in *The Ecological Indian: Myth and History* (New York: W. W. Norton, 1999).

4. Wilma A. Dunaway, "The Southern Fur Trade and the Incorporation of Southern Appalachia into the World Economy, 1690–1763," *Review (Fernand Braudel Center)* 17, no. 2 (Spring 1994): 215, 227, 230.

5. Ilo Hiller, *The White-Tailed Deer* (College Station: Texas A&M University Press, 1996), 94.

6. Thomas R. McCabe and Richard E. McCabe, "Recounting Whitetails Past," in *The Science of Overabundance: Deer Ecology and Population Management*, ed. William J. McShea, H. Brian Underwood, and John H. Rappel (Washington, D.C.: Smithsonian Institution Press, 1997), 15; Richard E. McCabe and Thomas R. McCabe, "Of Slings and Arrows: An Historical Retrospection," in *White-Tailed Deer: Ecology and Management*, ed. Lowell K. Halls (Harrisburg, Pa.: Stackpole Books, 1984), 30; David G. Hewitt ed., *Biology and Management of White-tailed Deer* (Boca Raton, Fla.: CRC Press, 2011), 357; Walter P. Taylor, ed., *The Deer of North America: The White-tailed, Mule and Black-tailed Deer, Genus Odocoileus, Their History and Management* (Harrisburg, Pa.: Stackpole, 1956), 2.

7. McCabe and McCabe, "Of Slings and Arrows," 20–21.

8. George A. Feldhamer, "Acorns and White-Tailed Deer: Interrelationships in Forest Ecosystems," in *Oak Forest Ecosystems: Ecology and Management for Wildlife*, ed. William J. McShea and William M. Healy (Baltimore: Johns Hopkins University Press, 2002): 221–22.

9. McCabe and McCabe, "Of Slings and Arrows," 28–29, 31–36; Hiller, *White-Tailed Deer*, 95; Hewitt, *Biology and Management of White-tailed Deer*, 356.

10. McCabe and McCabe, "Recounting Whitetails Past," 13.

11. McCabe and McCabe, "Of Slings and Arrows," 21, 39, 51; Hewitt, *Biology and Management of White-tailed Deer*, 358, 384.

12. G. Keith Parker, *Seven Cherokee Myths: Creation, Fire, the Primordial Parents, the Nature of Evil, the Family, Universal Suffering, and Communal Obligation* (Jefferson, N.C.: McFarland, 2006), 59–60; Robert J. Conley, *The Cherokee Nation: A History* (Albuquerque: University of New Mexico Press, 2005), 9; Krech, *Ecological Indian*, 168–70.

13. Lapham, *Hunting for Hides*, 3–5; William L. Anderson, Jane L. Brown, and Anne F. Rogers, eds., *The Payne-Butrick Papers* (Lincoln: University of Nebraska Press, 2010), 1:124–28, 234–35, quotes on 234.

14. "De Brahm's Account (1756)," in *Early Travels in the Tennessee Country, 1540–1800*, ed. Samuel Cole Williams (Johnson City, Tenn.: Watauga, 1928), 193.

15. Henry Timberlake, *The Memoirs of Lieut. Henry Timberlake* (London: J. Ridley, W. Nicoll, and C. Henderson, 1765), 42–45; Larry R. Kimball, "Early Archaic Settlement and Technology: Lessons from Tellico," in *The Paleoindian and Early Archaic Southeast*, ed. David G. Anderson and Kenneth E. Sassaman (Tuscaloosa: University of Alabama Press, 1996): 151–54, quote on 153.

16. James Adair quoted in Theda Perdue, *Slavery and the Evolution of Cherokee Society, 1540–1866* (Knoxville: University of Tennessee Press, 1979), 21–22.

17. Kimball, "Early Archaic Settlement and Technology," 159, 178; John H. DeWitt, "Old Fort Loudon," *Tennessee Historical Magazine* 3, no. 4 (December 1917): 250.

18. Jon Bernard Marcoux, *Pox, Empire, Shackles, and Hides: The Townsend Site, 1670–1715* (Tuscaloosa: University of Alabama Press, 2010), 38–40; W. Neil Franklin, "Virginia and the Cherokee Indian Trade, 1673–1752," *East Tennessee Historical Society Publications* 4 (January 1932): 4–6; Donald Davis, *Where There Are Mountains: An Environmental History*

of the Southern Mountains (Athens: University of Georgia Press, 2000), 62; Conley, *Cherokee Nation*, 21–22.

19. Christopher R. Rodning, *Center Places and Cherokee Towns: Archaeological Perspectives on Native American Architecture and Landscape in the Southern Appalachians* (Tuscaloosa: University of Alabama Press, 2015), 4; Tom Hatley, *Dividing Paths: Cherokees and South Carolinians through the Era of Revolution* (New York: Oxford University Press, 1993), 32; Marcoux, *Pox, Empire, Shackles, and Hides*, 12, 38–40.

20. Marcoux, *Pox, Empire, Shackles, and Hides*, 33–34; Dunaway, "Southern Fur Trade," 229–30; Franklin, "Virginia and the Cherokee Indian Trade, 1673–1752," 9; Paul C. Phillips, *The Fur Trade* (Norman: University of Oklahoma Press, 1961), 1:419–20; George Chicken, "Journal of Colonel George Chicken's Mission from Charleston, SC, to the Cherokees, 1725," in *Travels in the American Colonies*, ed. Newton D. Mereness (New York: Macmillan, 1916), 95; Christopher French, "Journal of an Expedition to South Carolina (1760–61)," *Journal of Cherokee Studies* 2 (Summer 1977): 276–77.

21. Perdue, *Slavery and the Evolution of Cherokee Society*, 21; Chicken, "Journal of Colonel George Chicken's Mission," 95; Philip M. Brown, "Early Indian Trade in the Development of South Carolina: Politics, Economic and Social Mobility during the Proprietary Period, 1670–1719," *South Carolina Historical Magazine* 76, no, 3 (July 1975): 122–23.

22. W. Matt Knox, "Historical Changes in the Abundance and Distribution of Deer in Virginia," in *The Science of Overabundance: Deer Ecology and Population Management*, ed. William J. McShea, H. Brian Underwood, and John H. Rappel (Washington, D.C.: Smithsonian Institution Press, 1997), 28; Brown, "Early Indian Trade in the Development of South Carolina," 122, 124; Dunaway, "Southern Fur Trade," 225; Marcoux, *Pox, Empire, Shackles, and Hides*, 33–34; Lapham, *Hunting for Hides*, 6–7.

23. Davis, *Where There Are Mountains*, 65.

24. George Chicken, "Journal of the March of the Carolinians into the Cherokee Mountain, in the Yemassee Indian War, 1715–1716," in *Charleston, SC, Yearbook, 1894*, ed. Langdon Cheves (Charleston: Walker, Evans, and Cogswell, 1895), 316, 347; Chicken, "Journal of Colonel George Chicken's Mission," 96; Phillips, *Fur Trade*, 1:419–20.

25. "Journal of Sir Alexander Cuming (1730)," in *Early Travels in the Tennessee Country, 1540–1800*, ed. Samuel Cole Williams (Johnson City, Tenn.: Watauga, 1928), 124–25, 138–41; Conley, *Cherokee Nation*, 28–29.

26. Taylor, *Deer of North America*, 16.

27. Krech, *Ecological Indian*, 156–57; Dunaway, "Southern Fur Trade," 226; David Wilcox, "Skins, Rum and Ruin: The Impact of the Colonial Deerskin Trade on the Southern Tribes, 1708–1782," *Southern Exposure* 13, no. 6 (November 1985): 58.

28. Hiller, *White-Tailed Deer*, 96; Lapham, *Hunting for Hides*, 10–11.

29. Phillips, *Fur Trade*, 1:404–5; Davis, *Where There Are Mountains*, 64.

30. Chicken, "Journal of the March of the Carolinians," 347; Marcoux, *Pox, Empire, Shackles, and Hides*, 12; Conley, *Cherokee Nation*, 40–41; McCabe and McCabe, "Of Slings and Arrows," 55.

31. Chahta Immataha quoted in Richard White, *The Roots of Dependency, Subsistence, Environment, and Social Change among the Choctaws, Pawnees, and Navajos* (Lincoln: University of Nebraska Press, 1983), 44.

32. "Journal of Antoine Bonnefoy, 1741–1742," in *Travels in the American Colonies*, ed. Newton D. Mereness (New York: Macmillan, 1916): 250; McCabe and McCabe, "Of

Slings and Arrows," 55; Wilcox, "Skins, Rum, and Ruin," 58; Dunaway, "The Southern Fur Trade," 225; Oliphant, *Peace and War on the Anglo-Cherokee Frontier*, 18–20.

33. Dunaway, "Southern Fur Trade," 227–28; Wilcox, "Skins, Rum, and Ruin," 59–60.

34. Wilcox, "Skins, Rum, and Ruin," 60.

35. "Bro. Martin Schneider's Report of His Journey to the Upper Cherokee Towns (1783–1784)," in Samuel Cole Williams, ed., *Early Travels in the Tennessee Country, 1540–1800* (Johnson City, Tenn.: Watauga, 1928), 253; Hiller, *White-Tailed Deer*, 97.

36. Wilcox, "Skins, Rum, and Ruin," 58; Brown, "Early Indian Trade in the Development of South Carolina," 121.

37. Lapham, *Hunting for Hides*, 10–11; Chicken, "Journal of Colonel George Chicken's Mission," 107, 129–34, quotes on 107 and 130.

38. Antoine Bonnefoy, "Journal of Antoine Bonnefoy, 1741–1742," in *Travels in the American Colonies*, ed. Newton D. Mereness (New York: Macmillan, 1916), 250; Oliphant, *Peace and War on the Anglo-Cherokee Frontier*, 35–36.

39. Davis, *Where There Are Mountains*, 67.

40. Hiller, *White-Tailed Deer*, 97; Philip M. Brown, "Early Indian Trade in the Development of South Carolina: Politics, Economic and Social Mobility during the Proprietary Period, 1670–1719," *South Carolina Historical Magazine* 76, no. 3 (July 1975): 123–24; Dunaway, "Southern Fur Trade," 227–28.

41. Lapham, *Hunting for Hides*, 9–10, 12; Hatley, *Dividing Paths*, 163–64.

42. Franklin, "Virginia and the Cherokee Indian Trade, 1673–1752," 7–8; Dunaway, "Southern Fur Trade," 229–30.

43. Brown, "Early Indian Trade in the Development of South Carolina," 118, 125, 127–28.

44. Sven Beckert, "Industrial Capitalism Takes Wing," chap. 6 in *Empire of Cotton: A Global History* (New York: Alfred A. Knopf, 2014).

45. Phillips, *Fur Trade*, 1:424–27; Paul M. Pressly, *On the Rim of the Caribbean: Colonial Georgia and the British Atlantic World* (Athens: University of Georgia Press, 2013), 17–19; Dunaway, "Southern Fur Trade," 227.

46. Taylor, *Deer of North America*, 23.

47. Rodning, *Center Places and Cherokee Towns*, 141, 158; Davis, *Where There Are Mountains*, 66; Conley, *Cherokee Nation*, 40–41.

48. Oliphant, *Peace and War on the Anglo-Cherokee Frontier*, 35; Taylor, *Deer of North America*, 23.

49. Hatley, *Dividing Paths*, 70.

50. Conley, *Cherokee Nation*, 45–48; Hatley, *Dividing Paths*, 95–96; Joe P. Distretti and Carl Kuttruff, "Reconstruction, Interpretation, and Education at Fort Loudon," in *The Reconstructed Past: Reconstructions in the Public Interpretation of Archeology and History*, ed. John H. Jameson Jr. (Walnut Creek, Calif.: AltaMira Press, 2004), 168; Franklin, "Virginia and the Cherokee Indian Trade, 1673–1752," 25.

51. "De Brahm's Account (1756)," in *Early Travels in the Tennessee Country, 1540–1800*, ed. Samuel Cole Williams (Johnson City, Tenn.: Watauga, 1928), 187, 189–90.

52. Conley, *Cherokee Nation*, 45–50; DeWitt, "Old Fort Loudon," 251–55; Perdue, *Slavery and the Evolution of Cherokee Society*, 34; Distretti and Kuttruff, "Reconstruction, Interpretation, and Education," 168.

53. French, "Journal of an Expedition to South Carolina," 284–89; Henry Timberlake, *The Memoirs of Lieut. Henry Timberlake* (London: J. Ridley, W. Nicoll, and C. Henderson, 1765).

54. William Bartram, *The Travels of William Bartram*, naturalist's ed., ed. Francis Harper (1958; repr., Athens: University of Georgia Press, 1998), 211, 217, 220, 231, 233.

55. Taylor, *Deer of North America*, 22; Phillips, *Fur Trade*, 1:411, 414; Krech, *Ecological Indian*, 163.

56. White, *Roots of Dependency*, 65–68; McCabe and McCabe, "Of Slings and Arrows," 25.

57. Timberlake, *Memoirs of Lieut. Henry Timberlake*, 19.

58. French, "Journal of an Expedition to South Carolina," 280.

59. W. Neil Franklin, "Virginia and the Cherokee Indian Trade, 1753–1775," *East Tennessee Historical Society Publications*, 5 (January 1933): 34–36; Hatley, *Dividing Paths*, 163–66, 212; Timberlake, *Memoirs of Lieut. Henry Timberlake*, 74; Phillips, *Fur Trade*, 546.

60. Timberlake, *Memoirs of Lieut. Henry Timberlake*, 76.

61. Hatley, *Dividing Paths*, 166.

62. William Bartram, "Some Hints & Observations, concerning the Civilization, of the Indians, or Aborigines of America," in *William Bartram: The Search for Nature's Design*, ed. Thomas Hillock and Nancy E. Hoffman (Athens: University of Georgia Press, 2010), 367; Oliphant, *Peace and War on the Anglo-Cherokee Frontier*, 18–20; Davis, *Where There Are Mountains*, 67–68, 73–76; Franklin, "Virginia and the Cherokee, 1753–1775," 35.

63. "Bro. Martin Schneider's Report of His Journey," 250, 253; David C. Hsiung, *Two Worlds in the Tennessee Mountains: Exploring the Origins of Appalachian Stereotypes* (Lexington: University Press of Kentucky, 1997), 83–84; Phillips, *Fur Trade*, 1:546; Hiller, *White-Tailed Deer*, 96; Davis, *Where There Are Mountains*, 66–67; Franklin, "Virginia and the Cherokee Indian Trade, 1753–1775," 38.

64. Charles Egbert Craddock, *The Story of Old Fort Loudon* (New York: Macmillan, 1899).

65. Distretti and Kuttruff, "Reconstruction, Interpretation, and Education," 169, 174.

66. Ibid., 169; William Bruce Wheeler and Michael J. McDonald, *TVA and the Tellico Dam, 1936–1979: A Bureaucratic Crisis in Post-Industrial America* (Knoxville: University of Tennessee Press, 1986), 148–55. For a detailed look at opposition to the Tellico Dam, see Zygmunt Plater, *The Snail Darter and the Dam: How Pork-Barrel Politics Endangered a Little Fish and Killed a River* (New Haven, Conn.: Yale University Press, 2013), 47; and Kenneth M. Murchison, *The Snail Darter Case: TVA versus the Endangered Species Act* (Lawrence: University Press of Kansas, 1994).

67. Plater, *Snail Darter and the Dam*, 338–41; Wheeler and McDonald, *TVA and the Tellico Dam*, 215; Distretti and Kattruff, "Reconstruction, Interpretation, and Education," 169, quote on 173.

68. Wheeler and McDonald, *TVA and the Tellico Dam*, 134.

69. DeWitt, "Old Fort Loudon," 256.

Chapter 2. Plants: Botanical Collectors and the Roots of Appalachian Identity

1. For Lyon's journal account of this summer, see Joseph Ewan and Nesta Ewan, "John Lyon, Nurseryman and Plant Hunter, and His Journal, 1799–1814," *Transactions of the American Philosophical Society* 53, no. 2 (1963): 27–29 (hereafter cited as Lyon's Journal).

Lyon described the offending plant as *Rhus pumila*, which was the contemporary name for Michaux's sumac (now *Rhus michauxii*). Michaux had described the nontoxic plant—which does not grow in the mountains—only the previous year, and Lyon likely had never encountered a specimen in the field. *Rhus vernix*, poison sumac, is relatively rare in the mountains, and this rarity coupled with Lyon's unfamiliarity with Michaux's sumac likely led to the misidentification. On the two species and their habits, see Albert E. Radford, Harry E. Ahles, and C. Ritchie Bell, *Manual of the Vascular Flora of the Carolinas* (Chapel Hill: University of North Carolina Press, 1968), 676, 678.

2. William Goodell Frost, "Our Contemporary Ancestors in the Southern Mountains," *Atlantic Monthly*, March 1899, 311–19; Ellen Churchill Semple, "The Anglo-Saxons of the Kentucky Mountains: A Study in Anthropogeography," *Bulletin of the American Geographical Society* 42, no. 8 (1910): 561–94. On the "discovery" of Appalachia, good starting points are Allen W. Batteau, *The Invention of Appalachia* (Tucson: University of Arizona Press, 1990); Henry D. Shapiro, *Appalachia on Our Mind: The Southern Mountains and Mountaineers in the American Consciousness, 1870–1920* (Chapel Hill: University of North Carolina Press, 1986); and David Whisnant, *All That Is Native and Fine: The Politics of Culture in an American Region* (Chapel Hill: University of North Carolina Press, 1984).

3. For some appearances of botanizers in histories of the region, see Timothy Silver, *Mount Mitchell and the Black Mountains: An Environmental History of the Highest Peaks in Eastern America* (Chapel Hill: University of North Carolina Press, 2003), 61–66; John Alexander Williams, *Appalachia: A History* (Chapel Hill: University of North Carolina Press, 2002), 83–87; Richard B. Drake, *A History of Appalachia* (Lexington: University Press of Kentucky, 2001), 120, 127; and Donald Edward Davis, *Where There Are Mountains: An Environmental History of the Southern Appalachians* (Athens: University of Georgia Press, 2000), 4, 44, 48, 54. Silver alone treats collectors in Appalachia as members of a profession in pursuit of commodities. Botanizers—evenly as radically circumscribed actors—sometimes all but disappear entirely from early Appalachian history. In *High Mountains Rising: Appalachia in Time and Place*, an excellent survey of the region edited by Richard A. Straw and H. Tyler Blethen, only William Bartram appears—and then in just a single sentence (Urbana: University of Illinois Press, 2004).

4. Drew Swanson, "Endangered Species and Threatened Habitats in Appalachia: Managing the Wild and the Human in the American Mountain South," *Environment and History* 18, no. 1 (February 2012): 44–45, 51–52; William Jackson Hooker, "On the Botany of America," *American Journal of Science and Arts* 9, no. 2 (January 2, 1825): 270; Asa Gray, *Scientific Papers of Asa Gray*, vol. 2, ed. Charles Sprague Sargent (Boston: Houghton, Mifflin, 1889). There are a number of historical and biographical studies of these botanists. See, for example, Leonard Warren, *Constantine Samuel Rafinesque: A Voice in the American Wilderness* (Lexington: University Press of Kentucky, 2005); A. Hunter Dupree, *Asa Gray: American Botanist, Friend of Darwin* (1959; repr., Baltimore: Johns Hopkins University Press, 1988); Henry Savage Jr. and Elizabeth J. Savage, *André and François André Michaux* (Charlottesville: University Press of Virginia, 1986); Edmund Berkeley and Dorothy Smith Berkeley, *A Yankee Botanist in the Carolinas: The Reverend Moses Ashley Curtis, D.D. (1808–1872)* (Berlin: J. Cramer, 1986); and Jeannette E. Graustein, *Thomas Nuttall, Naturalist: Explorations in America, 1808–1841* (Cambridge: Harvard University Press, 1967). The literature on the Bartrams (especially William) is too long to list here, but a starting place that connects father and son is Thomas P. Slaughter, *The Natures of John and William Bartram* (New York: Alfred A. Knopf, 1996). For an overview of the North American travels of

these and other botanists, see Ronald H. Peterson, *New World Botany: Columbus to Darwin* (Königstein, Germany: A. R. G. Gantner Verlag K. G., 2001). Chapter 9 deals most extensively with Appalachian journeys.

5. André Michaux, *Journal of André Michaux, 1793–1796*, Reuben Gold Thwaites, ed. (Cleveland: A. H. Clark, 1904), 55; Roy B. Clarkson, "Fraser's Sedge, *Cymophyllus fraseri* (Andrews) Mackenzie," *Castanea* 26, no. 4 (December 1961): 130; Ronald H. Peterson, "Moses Ashley Curtis's 1839 Expedition into the North Carolina Mountains," *Castanea* 53, no. 2 (June 1988): 113. On Curtis, see also Peterson, *New World Botany*, 443–461.

6. Lyon's Journal, 29.

7. Ibid.; François André Michaux, *Travels to the Westward of the Allegany Mountains, in the States of the Ohio, Kentucky, and Tennessee, in the Year 1802* (London: Barnard and Sultzer, 1805), 96; Michaux, *Journal of André Michaux, 1793–1796*, 56, 99.

8. Lyon's Journal, 8, 38; Peterson, "Moses Ashley Curtis's 1839 Expedition," 113; C. S. Rafinesque, *A Life of Travels and Researches in North America and South Europe* (Philadelphia: F. Turner, 1836), 71; Silver, *Mount Mitchell and the Black Mountains*, 63.

9. For a few studies of the evolution and expressions of the collecting impulse, see Fredrik Albritton Jonsson, "Climate Change and the Retreat of the Atlantic: The Cameralist Context of Pehr Kalm's Voyage to North America, 1748–51," *William and Mary Quarterly*, 3rd ser., 72, no. 1 (January 2015): 99–126; Anya Zilberstein, "Inured to Empire: Wild Rice and Climate Change," *William and Mary Quarterly*, 3rd ser., 72, no. 1 (January 2015): 127–58; Beth Fowkes Tobin, *The Duchess's Shells: Natural History Collecting in the Age of Cook's Voyages* (New Haven: Yale University Press, 2014); Londa Schiebinger, *Plants and Empire: Colonial Bioprospecting in the Atlantic World* (Cambridge: Harvard University Press, 2004); Richard Drayton, *Nature's Government: Science, Imperial Britain, and the 'Improvement' of the World* (New Haven: Yale University Press, 2000); Paula Findlen, *Possessing Nature: Museums, Collecting, and Scientific Culture in Early Modern Italy* (Berkeley: University of California Press, 1994); Oliver Impey and Arthur MacGregor eds., *The Origins of Museums: The Cabinet of Curiosities in Sixteenth- and Seventeenth-Century Europe* (Oxford: Clarendon Press, 1985); and Jay Tribby, "Body/Building: Living the Museum Life in Early Modern Europe," *Rhetorica* 10 (1992): 139–63.

10. Mark V. Barrow Jr., *Nature's Ghosts: Confronting Extinction from the Age of Jefferson to the Age of Ecology* (Chicago: University of Chicago Press, 2009), 48.

11. Drayton, *Nature's Government*, esp. chaps. 4–6; Ray Desmong, *Kew: A History* (London: Harvill Press, 1998).

12. Thomas Hallock and Nancy E. Hoffmann, eds., *William Bartram: The Search for Nature's Design* (Athens: University of Georgia Press, 2010), 83; Slaughter, *The Natures of John and William Bartram*, 172–73.

13. On Lyon's position in Philadelphia, see Asa Gray, *Scientific Papers of Asa Gray*, vol. 2, ed. Charles Sprague Sargent (Boston: Houghton, Mifflin, 1889), 32.

14. Lyon's Journal, 50–51.

15. Hooker, "On the Botany of America," 272.

16. Lyon's Journal, 50–51.

17. Bill Bryson, *A Short History of Nearly Everything* (New York: Broadway, 2003), 355.

18. Rodney H. True, "François André Michaux, the Botanist and Explorer," *Proceedings of the American Philosophical Society* 78, no. 2 (December 1937): 314–17; William James Robbins, "French Botanists and the Flora of the Northeastern United States: J. G. Milbert

and Elias Durand," *Proceedings of the American Philosophical Society* 101, no. 4 (August 1957): 364; Gilbert Chinard, "André and Francois André Michaux and Their Predecessors: An Essay on Early Botanical Exchanges between America and France," *Proceedings of the American Philosophical Society* 101, no. 4 (August 1957): 351–52; "Some North American Botanists, III: André Michaux," *Botanical Gazette* 8, no. 3 (March 1883): 181.

19. James Britten, "Biographical Notes. XXI.—Frasers' Catalogues," *Journal of Botany, British and Foreign* 37 (1899): 485, 486.

20. James Britten, "Thomas Walter (1740?–88) and His Grass," *Journal of Botany, British and Foreign* 59 (1921): 72–73.

21. Charles F. Jenkins, "Asa Gray and His Quest for Shortia Galacifolia," *Arnoldia* 2, nos. 3–4 (April 10, 1942): 18.

22. For the arc of Rafinesque's career, see Rafinesque, *Life of Travels and Researches*; for an example of new Appalachian botanical discoveries described in the *Atlantic Journal*, see C. S. Rafinesque, "New Plants of the Alleghany Mts.," *Atlantic Journal, and Friend of Knowledge* 1, no. 4 (Winter 1832): 153–54.

23. Correspondence, Curtis (Moses) and Engelmann (George), 1840–1871, Engelmann Papers, Missouri Botanical Garden, digitized and available at Biodiversity Heritage Library, www.biodiversitylibrary.org/bibliography/68768#/summary. For Engelmann's correspondence with other botanists and natural historians, see Letterbook, Nuttall (Thomas) and Engelmann (George), 1841; Letterbook, Gray (Asa) and Engelmann (George), volume 1841, both also in Engelmann Papers, Missouri Botanical Garden, digitized and available at Biodiversity Heritage Library, www.biodiversitylibrary.org/.

24. For examples, see Lyon's Journal, 27, 33.

25. Ibid., 29.

26. Ibid., 38.

27. Ibid., 35. On the contemporary nature of the term "sublime," see William Cronon, "The Trouble with Wilderness; or, Getting Back to the Wrong Nature," in *Uncommon Ground: Rethinking the Human Place in Nature*, ed. William Cronon (New York: W. W. Norton, 1996), 73–76.

28. William Bartram, *The Travels of William Bartram*, naturalist's ed., ed. Francis Harper (1958; repr., Athens: University of Georgia Press, 1998), 217. For a representative descriptive section, see 212–16.

29. William Bartram to unknown, April 1775, in Hallock and Hoffmann, *William Bartram*, 113.

30. Rafinesque, quoted in Asa Gray, *Notice of the Botanical Writings of the Late C. S. Rafinesque* (New Haven, Conn.: S. Converse, 1841), 222.

31. André Michaux, "Portions of the Journal of André Michaux, Botanist, Written during His Travels in the United States and Canada, 1785 to 1796," ed. Charles S. Sargent, *Proceedings of the American Philosophical Society* 26, no. 129 (January–July 1889): 112.

32. Garden quoted in Hooker, "On the Botany of America," 265.

33. François André Michaux, *Travels to the Westward*, 21, 23.

34. For example, see André Michaux, *Journal*, 45–49, 58.

35. Rafinesque, quoted in Gray, *Notice of the Botanical Writings*, 223–24.

36. Ibid., 224.

37. "Memoirs of the Life and Botanical Travels of André Michaux," *Belfast Monthly Magazine* 5, no. 24 (July 31, 1810): 41.

38. Hooker, "On the Botany of America," 267.

39. "Memoirs of the Life and Botanical Travels of André Michaux," 38–39; "Memoirs of the Life and Botanical Travels of André Michaux (continued)," *Belfast Monthly Magazine* 5, no. 25 (August 31, 1810): 117–18, 118–21, 123, quote on 125.

40. "Memoirs of the Life and Botanical Travels of André Michaux (continued)," 118.

41. Lyon's Journal, 35.

42. Gray, *Scientific Papers of Asa Gray*, 31.

43. Curtis quoted in Peterson, "Moses Ashley Curtis's 1839 Expedition," 114.

44. André Michaux, *Journal*, 45.

45. Rafinesque, *Life of Travels and Researches*, 71.

46. Robert Zahner and Steven M. Jones, "Resolving the Type Location for *Shortia galacifolia* T. & G.," *Castanea* 48, no. 3 (September 1983): 169; V. E. Vivian, "*Shortia galacifolia*: Its Life History and Microclimatic Requirements," *Bulletin of the Torrey Botanical Club* 94, no. 5 (September–October 1967): 369; P. A. Davies, "Type Location of *Shortia Galacifolia*," *Castanea* 21, no. 3 (September 1956): 107.

47. Davies, "Type Location of *Shortia Galacifolia*," 107.

48. Jenkins, "Asa Gray and His Quest," 13–28; Vivian, "*Shortia galacifolia*," 369; Davies, "Type Location of *Shortia Galacifolia*," 107; Asa Gray, "*Shortia Galacifolia* Re-discovered," *Botanical Gazette* 4, no. 1 (January 1879): 106.

49. J. W. Congdon, "Rediscovery of Shortia," *Bulletin of the Torrey Botanical Club* 6, no. 46 (October 1878): 266; Gray, "*Shortia Galacifloia* Re-discovered," 107–8.

50. Zahner and Jones, "Resolving the Type Location for *Shortia galacifolia*," 172; Davies, "Type Location of *Shortia Galacifolia*," 108.

51. "1810–Asa Gray–1885," *Botanical Gazette* 10, no. 8 (December 1885): 406–8.

52. C. S. Rafinesque, "Alleghanies Mountains," *Atlantic Journal, and Friend of Knowledge* 1, no. 5 (Spring 1833): 157–61.

53. Lyon's Journal, 27.

54. Ibid., 34. See Tiya Miles, *The House on Diamond Hill: A Cherokee Plantation Story* (Chapel Hill: University of North Carolina Press, 2010) on James Vann and the Moravian mission.

55. Lyon's Journal, 35.

56. For these examples, see ibid., 29 (quote), 48.

57. Jenkins, "Asa Gray and His Quest," 16.

58. Michaux, *Travels to the Westward*, 22, 27, 96.

59. Ibid., 38–39, quote on 45.

60. Elisha Mitchell, *Diary of a Geological Tour by Dr. Elisha Mitchell in 1827 and 1828* (Chapel Hill: University of North Carolina, 1905), 20. For an extended analysis of Mitchell's work in the mountains, see Timothy Silver, "Mitchell's Mountain," chap. 3 in *Mount Mitchell and the Black Mountains*.

61. Mitchell, *Diary of a Geological Tour*, 15.

62. Ibid., 18.

63. Ibid., 32–34.

64. Thomas Nuttall, "A Journal of Travels into the Arkansa Territory, during the Year 1819," in Reuben Gold Thwaites, ed., *Early Western Travels, 1748–1846*, vol. 13 (Cleveland: Arthur H. Clark, 1905), 58–59.

65. Ibid., 42.

66. Rafinesque, *Life of Travels and Researches*, 70.

67. Gray, *Scientific Papers of Asa Gray*, 45–46.

68. Jane Loring Gray, ed., *Letters of Asa Gray*, vol. 1 (London: MacMillan, 1893), 309.

69. William Bartram, *The Travels of William Bartram*, naturalist's ed., ed. Francis Harper (1958; repr., Athens: University of Georgia Press, 1998), xxiii, xxv.

70. André Michaux, *Quercus, or Oaks*, trans. Walter Wade (Dublin, Ireland: Graisberry and Campbell, 1809), 13.

71. "Memoirs of the Life and Botanical Travels of André Michaux," 36–42; "Memoirs of the Life and Botanical Travels of André Michaux (continued)," 117–27.

72. François André Michaux, *Travels to the Westward*.

73. Gray, *Scientific Papers*, vol. 2.

74. Ralph H. Lutts, "Like Manna from God: The American Chestnut Trade in Southwestern Virginia," *Environmental History* 9, no. 3 (July 2004): 497–525.

75. Kristin Johannsen, *Ginseng Dreams: The Secret World of America's Most Valuable Plant* (Lexington: University Press of Kentucky, 2006); David A. Taylor, *Ginseng, the Divine Root: The Curious History of the Plant that Captivated the World* (Chapel Hill, N.C.: Algonquin Books, 2006); Luke Manget, "Nature's Emporium: The Botanical Drug Trade and the Commons Tradition in Southern Appalachia, 1847–1917," *Environmental History* 21, no. 4 (October 2016): 660–87; Luke Manget, "Sangin' in the Mountains: The Ginseng Economy of the Southern Appalachians, 1865–1900," *Appalachian Journal* 40, nos. 1–2 (Fall 2012/Winter 2013): 28–55; Durwood Dunn, *Cades Cove: The Life and Death of a Southern Appalachian Community, 1818–1937* (Knoxville: University of Tennessee Press, 1988), 32–34. On the overall economic value of Appalachian forest products to local residents, see Kathy Newfont, "'The Custom of Our Country': The Appalachian Forest Commons," chap. 1 in *Blue Ridge Commons: Environmental Activism and Forest History in Western North Carolina* (Athens: University of Georgia Press, 2012); Davis, *Where There Are Mountains*, 106–16, 133–34, 144–46.

76. Will Wallace Harney, "A Strange Land and a Peculiar People," *Lippincott's Magazine of Popular Literature and Science* 12, no. 31 (October 1873): 429–38.

77. Michaux, "Portions of the Journal of André Michaux," 1–145; André Michaux, *Journal of André Michaux, 1787–1796* (Cambridge: Harvard University, 1888).

78. Asa Gray, *Scientific Papers of Asa Gray*, 2 vols., ed. Charles Sprague Sargent (Boston: Houghton, Mifflin, 1889).

79. Reuben Gold Thwaites, ed., *Early Western Travels, 1748–1846*, 32 volumes (Cleveland, Ohio: A. H. Clark, 1904–1907).

80. Samuel Cole Williams, *Early Travelers in the Tennessee Country* (Johnson City, Tenn.: Watauga Press, 1928).

81. Margaret W. Morley, *The Carolina Mountains* (Boston: Houghton Mifflin, 1913), 309–11, 334–35, 349–50.

82. Horace Kephart, *Our Southern Highlanders* (New York: Outing Publishing, 1913), 54–56, quote on 55.

83. Shepherd Dugger, *The Balsam Groves of the Grandfather Mountain: A Tale of the Western North Carolina Mountains* (Philadelphia: J. B. Lippincott, 1892), 45.

84. Ibid., 46.

85. Ibid., 144–60.

86. Shepherd Dugger, *The Balsam Groves of the Grandfather Mountain: A Tale of the Western North Carolina Mountains* (Banner Elk, N.C.: n.p., 1907), 260–73.

87. David Hsiung, *Two Worlds in the Tennessee Mountains: Exploring the Origins of Appalachian Stereotypes* (Lexington: University Press of Kentucky, 1997); Wilma A. Dunaway,

The First American Frontier: Transition to Capitalism in Southern Appalachia, 1700–1860 (Chapel Hill: University of North Carolina Press, 1996). It is worth noting that none of the prominent botanists of the colonial era or the early Republic appear in Dunaway's study of regional economics, despite her emphasis on the importance of markets and commodities from the earliest days of Euro-American settlement.

88. Frost, "Our Contemporary Ancestors in the Southern Mountains," 311–19; Frederick Jackson Turner, "The Significance of the Frontier in American History," in *Proceedings of the Forty-First Annual Meeting of the State Historical Society of Wisconsin* (Madison: State Historical Society of Wisconsin, 1894).

Chapter 3. Gold: The Rise, Fall, and Rebirth of Southern Gold Mining

1. David Williams, *The Georgia Gold Rush: Twenty-Niners, Cherokees, and Gold Fever* (Columbia: University of South Carolina Press, 1993); Otis E. Young Jr., "The Southern Gold Rush, 1828–1836," *Journal of Southern History* 48, no. 3 (August 1982): 373–92; E. Merton Coulter, *Auraria: The Story of a Georgia Gold-Mining Town* (Athens: University of Georgia Press, 1956).

2. Stephen Mihm, *A Nation of Counterfeiters: Capitalists, Con Men, and the Making of the United States* (Cambridge: Harvard University Press, 2007).

3. Edward Baptist, *The Half Has Never Been Told: Slavery and the Making of American Capitalism* (New York: Basic Books, 2014), 90–92. In the Yazoo fraud, Georgia state officials sold land in today's Alabama and Mississippi through insider deals, the Supreme Court nullified the transactions, and the result was a legal quagmire of competing land claims that took years to resolve. For an introduction to the topic, see Charles F. Hobson, *The Great Yazoo Lands Sale: The Case of "Fletcher v. Peck"* (Lawrence: University Press of Kansas, 2016).

4. *Compendium of the Enumeration of the Inhabitants and Statistics of the United States* (Washington, D.C.: Thomas Allen, 1841), 204.

5. For recent studies of antebellum calls for economic diversity in Georgia, especially in the form of agricultural diversity, see William Thomas Okie, "'Everything Is Peaches Down in Georgia': Culture and Agriculture in the American South" (PhD diss., University of Georgia, 2012), 14–61; Drew Swanson, *Remaking Wormsloe Plantation: The Environmental History of a Lowcountry Landscape* (Athens: University of Georgia Press, 2012), chap. 2.

6. Fletcher Green pioneered the historical study of the state's gold rush in a pair of articles in the *Georgia Historical Quarterly*: Green, "Georgia's Forgotten Industry: Gold Mining, Part I," *Georgia Historical Quarterly* 19, no. 2 (June 1935): 93–111; Green, "Georgia's Forgotten Industry: Gold Mining, Part II," *Georgia Historical Quarterly* 19, no. 3 (September 1935): 210–28. The most comprehensive treatment of the early gold rush is Williams, *Georgia Gold Rush*. Local historian Anne Dismukes Amerson has also compiled modern memories of people with a connection to mining center Dahlonega, Georgia, including many stories of the initial gold rush, printed in a series titled "I Remember Dahlonega." The first volume is *"I Remember Dahlonega": Memories of Growing Up in Lumpkin County as Told to Anne Dismukes Amerson* (Dahlonega, Ga.: Chestatee Publications, 1990).

7. Kate Brown, *Dispatches from Dystopia: Histories of Places Not Yet Forgotten* (Chicago: University of Chicago Press, 2015), 32. On the American love of origin myths, see David Nye, *America as Second Creation: Technology and Narratives of New Beginnings* (Cambridge: Massachusetts Institute of Technology Press, 2004).

8. Williams, *Georgia Gold Rush*, 21–25; H. David Williams, "Origins of the North Georgia Gold Rush," *Proceedings of the Georgia Association of Historians* 9 (1988): 161; Young, "Southern Gold Rush," 375; Lou Harshaw, *Gold of Dahlonega: The First Major Gold Rush in North America* (Asheville, N.C.: Hexagon, 1976), 18; W. S. Yeates, S. W. McCallie, and Francis P. King, *A Preliminary Report on a Part of the Gold Deposits of Georgia*, Geological Survey of Georgia Bulletin 4-A (Atlanta: George W. Harrison, 1896), 271–73.

9. Williams, "Origins of the North Georgia Gold Rush," 162–64.

10. See "History of Villa Rica," www.pinemountaingoldmuseum.com/history.html.

11. Yeates, McCallie, and King, *Preliminary Report on a Part of the Gold Deposits*, 8.

12. Ibid., 162; Fletcher M. Green, "Gold Mining: A Forgotten Industry of Antebellum North Carolina," *North Carolina Historical Review* 14 (1937): 1–19, 135–55; Fletcher M. Green, "Gold Mining in Antebellum Virginia," *Virginia Magazine of History and Biography* 45 (1937): 227–35, 357–66; Denison Olmsted, "On the Gold Mines of North Carolina," *American Journal of Science and Arts* 9, no. 1 (January 1, 1825): 5–15.

13. Elisha Mitchell, *Diary of a Geological Tour by Dr. Elisha Mitchell in 1827 and 1828* (Chapel Hill: University of North Carolina, 1905), 43.

14. Young, "Southern Gold Rush," 384, 389–90.

15. Williams, *Georgia Gold Rush*, 25; Williams, "Origins of the North Georgia Gold Rush," quote on 165.

16. King quoted in Sherry L. Boatright, *The John C. Calhoun Gold Mine: An Introductory Report* (Atlanta: Department of Natural Resources, Historic Preservation Section, 1974), 6.

17. "The Georgia Gold Region," *Niles' Weekly Register* 44, no. 1128 (May 4, 1833): 152.

18. Williams, "'Prosper the Americans and Cherokees': The Climactic Year of 1838," chap. 7 in *Georgia Gold Rush*; Williams, "Origins of the North Georgia Gold Rush," 165–67; Coulter, *Auraria*, 4–5.

19. Wilma A. Dunaway, *The First American Frontier: Transition to Capitalism in Southern Appalachia, 1700–1860* (Chapel Hill: University of North Carolina Press, 1996), 169–70; Williams, *Georgia Gold Rush*, 47–55; David Williams, "Georgia's Forgotten Miners: African-Americans and the Georgia Gold Rush," *Georgia Historical Quarterly* 75, no. 1 (Spring 1991): 78; Boatright, *John C. Calhoun Gold Mine*, 9.

20. See, for example, the legal papers of a lawsuit of Johnson Ledbetter against Martin Roberts, Jeptha Scott, and Jonas Bedford, Lumpkin County Gold Scam Collection, ms 3458, Hargrett Rare Book and Manuscript Library, University of Georgia, Athens, Georgia (hereafter cited as Hargrett); Boatright, *John C. Calhoun Gold Mine*, 17–19.

21. William P. Blake and Charles T. Jackson, *Gold Placers of the Vicinity of Dahlonega, Georgia* (Boston, 1859), 4–7.

22. Young, "Southern Gold Rush," 377–78; Southern Gold Company, *Prospectus of the Southern Gold Company* (n.p, 1859), 9.

23. "On the Diminished Product and Probable Exhaustion of the Gold Mines of the United States," *Farmer's Register* 3, no. 12 (April 1836): 738–40.

24. Southern Gold Company, *Prospectus of the Southern Gold Company*, 6.

25. Williams, *Georgia Gold Rush*, 70–73; *Nacoochee Hydraulic Mining Company* (Boston, n.d.), 20.

26. Blake and Jackson, *Gold Placers of the Vicinity of Dahlonega*, 7.

27. Thomas G. Clemson, "Gold and the Gold Region," *Orion* 4, no. 2 (April 1844): 65.

28. Williams, *Georgia Gold Rush*, 70–73; Young, "Southern Gold Rush," 387–89; Coulter, *Auraria*, 11–13; "Dahlonega, or Georgia Gold Region," *Merchants' Magazine and Commercial Review* 19, no. 1 (July 1, 1848): 113; "The Georgia Gold Region," 152.

29. John Hamlin Newton Account Book, ms 228, Hargrett.

30. Wilma Dunaway, *Slavery in the American Mountain South* (Cambridge, UK: Cambridge University Press, 2003), 119–25; Williams, "Georgia's Forgotten Miners," 76–89; Young, "Southern Gold Rush," 381–82.

31. John C. Inscoe, *Mountain Masters: Slavery and the Sectional Crisis in Western North Carolina* (Knoxville: University of Tennessee Press, 1989), 71–73.

32. Williams, "Georgia's Forgotten Miners," 84–85.

33. Young, "Southern Gold Rush," 379–80.

34. Porte Crayon (David Hunter Strother), "North Carolina Illustrated: IV, The Gold Region," *Harper's New Monthly Magazine* 15, no. 87 (August 1857): 297–98.

35. Williams, *Georgia Gold Rush*, 77; Coulter, *Auraria*, 11–13.

36. "An Act to Incorporate the Augusta Mining Company, the Habersham Mining Company, and the Naucoochy Mining Company," *Acts of the General Assembly of the State of Georgia, 1831*, vol. 1 (Milledgeville, Ga.: Prince and Ragland, 1832), 88–94; "An Act to Incorporate the Pigeon-roost Mining Company and the Belfast Mining Company of Lumpkin County," *Acts of the General Assembly of the State of Georgia, 1834*, vol. 1 (Milledgeville, Ga.: P. L. and B. H. Robinson, 1835), 143–144; "An Act to Incorporate the Georgia Mining Company, the Chestatee Mining Company, and the Cherokee Mining Company," *Acts of the General Assembly of the State of Georgia, 1835*, vol. 1 (Milledgeville, Ga.: John A. Cuthbert, 1836), 128–31; "An Act to Incorporate Certain Persons, under the Name and Style of the Lumpkin County Mining and Manufacturing Company," *Acts of the General Assembly of the State of Georgia, 1837*, vol. 1 (Milledgeville, Ga.: P. L. Robinson, 1838), 130–31.

37. Young, "Southern Gold Rush," 373.

38. Williams, *Georgia Gold Rush*, 105–8; Sylvia Head and Elizabeth Etheridge, *The Neighborhood Mint: Dahlonega in the Age of Jackson* (Macon, Ga.: Mercer University Press, 1986); Claire Birdsall, *The United States Branch Mint at Dahlonega, Georgia: Its History and Coinage* (Easley, S.C.: Southern Historical Press, 1984); S. P. Jones, *Second Report on the Gold Deposits of Georgia*, Geological Survey of Georgia, Bulletin 19 (Atlanta: Chas. P. Byrd, 1909), 17.

39. Young, "Southern Gold Rush," 382; Coulter, *Auraria*, 23–24. Young concludes that the total amount of southern gold mined may have been larger than the federal estimates by as much as a factor of ten.

40. Yeates, McCallie, and King, *Preliminary Report on a Part of the Gold Deposits*, 271–73.

41. Boatright, *John C. Calhoun Gold Mine*, 12, 17–19, 32, 36, 40; Williams, "Georgia's Forgotten Miners," 79–80.

42. Coulter, *Auraria*, 8–9; Marielle Lumang, "There's Gold in Them Thar Hills!" (master's thesis, University of Georgia, 2007), 27–29.

43. "Dahlonega, or Georgia Gold Region," 112.

44. Blake and Jackson, *Gold Placers of the Vicinity of Dahlonega*, 6.

45. "Sketches of Georgia," *Southern Literary Messenger* 6, no. 11 (November 1840): 777.

46. For descriptions of the lawless nature of the early goldfields, see Williams, "'Gambling Houses, Dancing Houses, & Drinking Saloons': Life in the Georgia Gold Region," chap. 6 in *Georgia Gold Rush*; Coulter, *Auraria*.

47. "Sketches of Georgia," 776.

48. Clemson quoted in Boatright, *John C. Calhoun Gold Mine*, 47.

49. Clemson, "Gold and the Gold Region," 65.

50. Isaac Boring, "Mission in the Cherokee District," *Christian Advocate and Journal* 8, no. 42 (June 13, 1834): 166; H. P. Pitchford, "Chestatee Mission," *Christian Advocate and Journal* 9, no. 39 (May 22, 1835): 154.

51. "Interior Georgia Life and Scenery by a Southern Traveler," *Knickerbocker Magazine* 34 (August 1849): 113–18. Reprinted in *Travels in the Old South: Selected from Periodicals of the Times*, ed. Eugene L. Schwaab with Jacqueline Bull (Lexington: University Press of Kentucky, 1973), 2:414.

52. "Dahlonega, or Georgia Gold Region," 112; Lumang, "There's Gold in Them Thar Hills," 30–34.

53. William Gilmore Simms, *Guy Rivers: A Tale of Georgia* (New York: Harper and Brothers, 1834), 1–2.

54. Ibid., 3. John Inscoe argues that the Appalachian setting of *Guy Rivers* is of little relevance to the plot, as Simms created similar tales contrasting the detrimental influence of various frontier landscapes with the relative benefits of more settled regions. See Inscoe, "The 'Ferocious Character' of Antebellum Georgia's Gold Country: Frontier Lawlessness and Violence in Fact and Fiction," chap. 4 in *Blood in the Hills: A History of Violence in Appalachia*, ed. Bruce Stewart (Lexington: University Press of Kentucky, 2012).

55. Young, "Southern Gold Rush," 373, 383, 391.

56. "Dahlonega, or Georgia Gold Region," 112.

57. Williams, *Georgia Gold Rush*, 118–19; Williams, "Georgia's Forgotten Miners," 86; Coulter, *Auraria*, 1.

58. Young, "Southern Gold Rush," 391; Boatright, *John C. Calhoun Gold Mine*, 56.

59. Williams, *Georgia Gold Rush*, 4.

60. Coulter, *Auraria*, 111–12; Ben Green Cooper, *Dahlonega Gold* (Atlanta: Atlanta Coin, 1962), 12–13.

61. The best account of the techniques and destructiveness of California's antebellum mining industry is Andrew Isenberg, *Mining California: An Ecological History* (New York: Hill & Wang, 2005), and some interesting technical details can be found in Hunter Rouse, *Hydraulics in the United States, 1776–1976* (Iowa City, Iowa: Institute of Hydraulic Research, 1976), 50–54. For a contemporary description of the practice, see Blake and Jackson, *Gold Placers of the Vicinity of Dahlonega*, 11.

62. Robert Righter, *The Battle over Hetch Hetchy: America's Most Controversial Dam and the Birth of Modern Environmentalism* (New York: Oxford University Press, 2005), 32.

63. See "Malakoff Diggins State Historic Park," www.parks.ca.gov/?page_id=494. Paul Sutter's *Let Us Now Praise Famous Gullies: Providence Canyon and the Soils of the South* (Athens: University of Georgia Press, 2015), 205–8, discusses Malakoff Diggins as a site commemorating human-induced soil erosion.

64. "Patent Fish Hooks," *Scientific American* 3, no. 51 (September 8, 1848): 406; Anthropos, "American Contributions to Science," *United States Magazine* 4, no. 2 (February 1857): 173; David B. Dill, "William Phipps Blake, Yankee Gentleman and Pioneer Geologist of the Far West," *Journal of Arizona History* 32, no. 4 (Winter 1991): 385–412.

65. William P. Blake, "On the Grooving and Polishing of Hard Rocks and Minerals by Dry Sand," *Pioneer; or, California Monthly Magazine* 4, no. 2 (August 1855): 111.

66. William P. Blake, *The Published Writings of William Phipps Blake, D. Sc., LL. D.* (Tucson: University of Arizona, 1910); Dill, "William Phipps Blake," 385–412.

67. Blake and Jackson, *Gold Placers of the Vicinity of Dahlonega*, 10.

68. Ibid., 60–61. This particular endorsement came from John Blake, William's brother and an engineer.

69. William P. Blake, *Prospectus of the Chestatee Hydraulic Company, and Report upon the Gold Placers of a Part of Lumpkin County, Georgia, and the Practicality of Working Them by the Hydraulic Method, with Water from the Chestatee River* (New York: John F. Trow, 1858), 4.

70. "The American Association for the Advancement of Science," *Scientific American* 3, no. 9 (August 25, 1860): 133.

71. Blake and Jackson, *Gold Placers of the Vicinity of Dahlonega*; Blake, *Prospectus of the Chestatee Hydraulic Company*; William P. Blake, *Report upon the Property of the Mining Company Called the Auraria Mines of Georgia* (Boston, 1859); Southern Gold Company, *Prospectus of the Southern Gold Company*, 5; *Nacoochee Hydraulic Mining Company*, 6–7; Minutes of the Stephenson Gold Mining Company in the Sons of Temperance Ledger, ms 3096, Hargrett.

72. Amory Dexter Diary, 1861 (photocopy), ms 320, Georgia State Archives, Morrow.

73. "An Act to Incorporate the Loud Hydraulic Hose Mining Company," *Acts of the General Assembly of the State of Georgia, 1860* (Milledgeville, Ga.: Boughton, Lisbet & Barnes, 1861), 1:124–25; "An Act to Incorporate the Wood Hydraulic Hose Gold Mining Company," *Acts of the General Assembly of the State of Georgia, 1860*, 1:132–33; "An Act to Incorporate the Cavender's Creek and Field Mining Company," *Acts of the General Assembly of the State of Georgia, 1861* (Milledgeville, Ga.: Boughton, Lisbet & Barnes, 1862), 1:97–99.

74. Blake and Jackson, *Gold Placers of the Vicinity of Dahlonega*; Southern Gold Company, *Prospectus of the Southern Gold Company*, 12, 14.

75. Adelberg & Raymond Mining Reports, folders 1–3, 5, ms. 1355, Hargrett; Adelberg & Raymond Company, *Report on the Lewis Gold Mine, White County*, Georgia (New York: Stockholder Job Printing, 1866).

76. Randal L. Hall, *Mountains on the Market: Industry, the Environment, and the South* (Lexington: University Press of Kentucky, 2012), 68–72; Daniel Yergin, *The Prize: The Epic Quest for Oil, Money, and Power* (New York: Simon and Schuster, 1991), 21–22. On the wide-ranging importance of geological surveyors in identifying and promoting mineral deposits and other natural resources during the era, see Denise D. Meringolo, *Museums, Monuments, and National Parks: Toward a New Genealogy of Public History* (Amherst: University of Massachusetts Press, 2012), 22–25.

77. Benjamin Cohen, *Notes from the Ground: Science, Soil, and Society in the American Countryside* (New Haven, Conn.: Yale University Press, 2011); Steven Stoll, *Larding the Lean Earth: Soil and Society in Nineteenth-Century America* (New York: Hill & Wang, 2002); William Mathew, *Edmund Ruffin and the Crisis of Slavery in the Old South: The Failure of Agricultural Reform* (Athens: University of Georgia Press, 1988); Drew Swanson, *A Golden Weed: Tobacco and Environment in the Piedmont South* (New Haven, Conn.: Yale University Press, 2014), 82–118.

78. Blake and Jackson, *Gold Placers of the Vicinity of Dahlonega*, 11–12.

79. "Hydraulic Gold-Mining," *Scientific American* 1, no. 23 (December 3, 1859): 369.

80. Blake, *Prospectus of the Chestatee Hydraulic Company*, 5.

81. *Nacoochee Hydraulic Mining Company*, 5.

82. Blake and Jackson, *Gold Placers of the Vicinity of Dahlonega*, 33.

83. Ibid., 3.

84. Blake and Jackson, *Gold Placers of the Vicinity of Dahlonega*, 37–38.

85. Williams, *Georgia Gold Rush*, 120.

86. Blake, *Prospectus of the Chestatee Hydraulic Company*, 7.

87. Amory Dexter Diary, 1861.

88. Yeates, McCallie, and King, *Preliminary Report on a Part of the Gold Deposits*, 276.

89. There is evidence that some hydraulic mining took place during the war's initial months, including the organization of a new company with M. H. Van Dyke as chairman. See Sons of Temperance Ledger (which includes the minutes of the Stephenson Gold Mining Company from September and October, 1861, ms 3096, Hargrett.

90. "Report on the Mineral Property of Dr. B. Hamilton near Actworth, Ga., Formerly Called O'Neil's Gold Property, Oct 19, 1866," Adelberg & Raymond Mining Reports, folder 3, Hargrett.

91. Yeates, McCallie, and King, *Preliminary Report on a Part of the Gold Deposits*, 531.

92. Adelberg & Raymond Company, *Report on the Lewis Gold Mine, White County, Georgia*, 1–11; *Nacoochee Hydraulic Mining Company*, 3; Standard Gold Company, *Prospectus, Standard Gold Mining Company, Dahlonega, Georgia* (Dahlonega, Ga.: Standard Gold Company, n.d.), 3, 7, 19, 23; R. F. T., "Present Conditions of the South," *Advocate of Peace* 62, no. 3 (March 1900): 59–60; Williams, *Georgia Gold Rush*, 120–21. For turn-of-the-century mining scenes, see Johnston Family Papers, box 5, folder 3, Hargrett.

93. Kathryn Morse, *The Nature of Gold: An Environmental History of the Klondike Gold Rush* (Seattle: University of Washington Press, 2010); William Kelleher Storey, *Guns, Race, and Power in Colonial South Africa* (Cambridge, UK: Cambridge University Press, 2008), 321; Nancy Keesing, *History of the Australian Gold Rushes, by Those Who Were There* (New York: Harper Collins, 1987).

94. For an interesting recent study of the importance of nineteenth-century American metal mining, see Kent A. Curtis, *Gambling on Ore: The Nature of Metal Mining in the United States, 1860–1910* (Boulder: University Press of Colorado, 2013).

95. Hall, *Mountains on the Market*, 64–73; Duncan Maysilles, "The Setting, The Cherokees, and the First Era of Ducktown Mining, 1843–1878," chap. 1 in *Ducktown Smoke: The Fight over One of the South's Greatest Environmental Disasters* (Chapel Hill: University of North Carolina Press, 2011).

96. Chad Montrie, "Degrees of Separation," chap. 4 in *Making a Living: Work and Environment in the United States* (Chapel Hill: University of North Carolina Press, 2008); John Alexander Williams, *West Virginia and the Captains of Industry* (1976; repr., Morgantown: West Virginia University Press, 2003); Ronald Lewis, *Transforming the Appalachian Countryside: Railroads, Deforestation, and Social Change in West Virginia, 1880–1920* (Chapel Hill: University of North Carolina Press, 1998); Crandall Shifflett, *Coal Towns: Life, Work, and Culture in Company Towns of Southern Appalachia, 1880–1960* (Knoxville: University of Tennessee Press, 1991); Ronald Eller, *Miners, Millhands, and Mountaineers: Industrialization of the Appalachian South, 1880–1930* (Knoxville: University of Tennessee Press, 1982).

Chapter 4. Salt: Saltville's Civil War

1. J. B. Jones, *A Rebel War Clerk's Diary, at the Confederate States Capital* (Philadelphia: J. B. Lippincott, 1866), 1:183–84.

2. Ibid., 1:194.

3. Randolph H. McKim, *A Soldier's Recollections: Leaves from the Diary of a Young Confederate, with an Oration on the Motives and Aims of the Soldiers of the South* (New York: Longmans, Green, 1921), 79, 63.

4. Ella Lonn, *Salt as a Factor in the Confederacy* (Tuscaloosa: University of Alabama Press, 1965); Mark Kurlansky, "The War between the Salts," chap. 16 in *Salt: A World History* (New York: Penguin Books, 2003).

5. For selected examples of the scholarship on Appalachia's Civil War, see Steven E. Nash, "Setting the Stage: Antebellum and Civil War Western North Carolina," chap. 1 in *Reconstruction's Ragged Edge: The Politics of Postwar Life in the Southern Mountains* (Chapel Hill: University of North Carolina Press, 2016); Jonathan Dean Sarris, *A Separate Civil War: Communities in Conflict in the Mountain South* (Charlottesville: University of Virginia Press, 2006); Kenneth W. Noe and Shannon H. Wilson, *The Civil War in Appalachia: Collected Essays* (Knoxville: University of Tennessee Press, 2004); Phillip Shaw Paludan, *Victims: A True Story of the Civil War*, revised ed. (Knoxville: University of Tennessee Press, 2004); John W. Shaffer, *Clash of Loyalties: A Border County in the Civil War* (Morgantown: West Virginia University Press, 2003); Martin Crawford, *Ashe County's Civil War: Community and Society in the Appalachian South* (Charlottesville: University of Virginia Press, 2001); Noel C. Fisher, *War at Every Door: Partisan Politics and Guerrilla Violence in East Tennessee, 1860–1869* (Chapel Hill: University of North Carolina Press, 2001); John C. Inscoe and Gordon B. McKinney, *The Heart of Confederate Appalachia: Western North Carolina in the Civil War* (Chapel Hill: University of North Carolina Press, 2000); and W. Todd Groce, *Mountain Rebels: East Tennessee Confederates and the Civil War, 1860–1870* (Knoxville: University of Tennessee Press, 2000).

6. Wilma A. Dunaway, *The First American Frontier: Transition to Capitalism in Southern Appalachia, 1700–1860* (Chapel Hill: University of North Carolina Press, 1996), 227–29.

7. Randal L. Hall, *Mountains on the Market: Industry, the Environment, and the South* (Lexington: University Press of Kentucky, 2012), 5.

8. Maureen S. Meyers, "The Mississippian Frontier in Southwestern Virginia," *Southeastern Archeology* 21, no. 2 (Winter 2002): 188.

9. C. B. Hayden, "On the Rock Salt and Salines of the Holston," *American Journal of Science and Arts* 44, no. 1 (October–December 1842): 173.

10. Thomas Jefferson, *Notes on the State of Virginia*, ed. William Peden (New York: W. W. Norton, 1972), 34.

11. C. D., "Remarks, —Topographical, Geological, and General, Respecting Preston's and King's Saltworks, and the Surrounding District of Country," *Farmers' Register* 1, no. 8 (Jan 1834): 498–500; William H. Nicholls, "Some Foundations of Economic Development in the Upper East Tennessee Valley, 1850–1900," *Journal of Political Economy* 64, no. 4 (August 1956): 290.

12. E. Meriam, "Virginia Salt Mine," *Farmers' Register* 10, no. 1 (January 31, 1842): 21; Hayden, "On the Rock Salt and Salines," 174–75; Charles Rufus Boyd, *Resources of South-West Virginia* (New York: John Wiley & Sons, 1881), 102–3; Lonn, *Salt as a Factor in the Confederacy*, 25–26.

13. Hayden, "On the Rock Salt and Salines," 173–74, 177–78; C. D., "Remarks, —Topographical, Geological, and General," 499–500.

14. Hayden, "On the Rock Salt and Salines, 175, 177; C. D., "Remarks, —Topographical, Geological, and General," 500.

15. Wilma Dunaway, *Slavery in the American Mountain South* (Cambridge, UK: Cambridge University Press, 2003), 117–19; John Alexander Williams, *Appalachia: A History* (Chapel Hill: University of North Carolina Press, 2002), 129; Ralph Mann, "Mountains, Land, and Kin Networks: Burkes Garden, Virginia, in the 1840s and 1850s," *Journal of Southern History* 58, no. 3 (August 1992): 419.

16. Dunaway, *Slavery in the American Mountain South*, 118; Booker T. Washington, *Up from Slavery: An Autobiography* (New York: Doubleday, Page, 1906), 24–27.

17. Frederick Law Olmsted, *A Journey in the Back Country* (New York: Mason Brothers, 1860), 273–74; Hall, *Mountains on the Market*, 75–80.

18. Nicholls, "Some Foundations of Economic Development," 290.

19. C. D., "Remarks, —Topographical, Geological, and General," 498.

20. "Trade and Manufacture of Salt in the United States," *Merchants' Magazine and Commercial Review* 8, no. 4 (April 1, 1843): 357–64.

21. Angus J. Johnston II, "Virginia Railroads in April 1861," *Virginia Magazine of History and Biography* 22, no. 3 (August 1957): 312, 319; Kenneth W. Noe, *Southwest Virginia's Railroad: Modernization and the Sectional Crisis in the Civil War Era* (Tuscaloosa: University of Alabama Press, 2003); David C. Hsiung, *Two Worlds in the Tennessee Mountains: Exploring the Origins of Appalachian Stereotypes* (Lexington: University Press of Kentucky, 1997), 133–55.

22. C. D., "Remarks, —Topographical, Geological, and General," 500. Peak production was in 1839, according to Nicholls, "Some Foundations of Economic Development," 291.

23. Will Sarvis, "An Appalachian Forest: Creation of the Jefferson National Forest and Its Effects on the Local Community," *Forest and Conservation History* 37, no. 4 (October 1993): 177n14; Donald Edward Davis, *Where There Are Mountains: An Environmental History of the Southern Appalachians* (Athens: University of Georgia Press, 2000), 154–55; Lonn, *Salt as a Factor in the Confederacy*, 121–22.

24. On other regional mining and industrial activities, see Hall, *Mountains on the Market*.

25. C. D., "Remarks, —Topographical, Geological, and General," 498.

26. Frederick Jackson Turner, *The Frontier in American History* (New York: Henry Holt, 1921), 18.

27. Dunaway, *Slavery in the American Mountain South*, 117; Dunaway, *First American Frontier*, 315.

28. On Confederate ideas about agricultural production and its relationship to military might, see R. Douglas Hurt, *Agriculture and the Confederacy: Policy, Productivity, and Power in the Civil War South* (Chapel Hill: University of North Carolina Press, 2015). For regional flows of northern foodstuffs into the Deep South (as well as an argument against overstating the South's dependency on Midwestern pork), see Sam Bowers Hilliard, *Hog Meat and Hoecake: Food Supply in the Old South, 1840–1860* (1972; repr., Athens: University of Georgia Press, 2014).

29. Lonn, *Salt as a Factor in the Confederacy*, 190–93; James M. McPherson, *Battle Cry of Freedom: The Civil War Era* (New York: Ballantine Books, 1988), 440.

30. James M. McPherson, *Embattled Rebel: Jefferson Davis and the Confederate Civil War* (New York: Penguin, 2014), 166–67.

31. *The War of the Rebellion: A Compilation of the Official Records of the Union and Confederate Armies* (hereafter cited as *O.R.*), series 4, vol. 1 (Washington: GPO, 1900), 872–78, quote on 878.

32. *O.R.*, series 1, vol. 52, pt. 2, 383.

33. Ibid., 384.

34. Mary Boykin Chestnut, *A Diary from Dixie*, ed. Isabella D. Martin and Myrta Lockett (New York: D. Appleton, 1905), 261.

35. *O.R.*, series 4, vol. 2, 181–82; J. G. De Roulhac Hamilton, *Reconstruction in North Carolina* (1914; repr., New York: Columbia University Press, 1964), 75–76.

36. *O.R.*, series 4, vol. 2, 182; Lonn, *Salt as a Factor in the Confederacy*, 93–97, 104–5.

37. Paludan, *Victims*, 81–83.

38. Reeves quoted in Crawford, *Ashe County's Civil War*, 106.

39. *O.R.*, series 1, vol. 51, pt. 2, 1056–57, 1061–62, quote on 1047.

40. *O.R.*, series 4, vol. 3, 924–25.

41. *O.R.*, series 1, vol. 51, pt. 2, 1058–62.

42. *O.R.*, series 4, vol. 3, 925; Lonn, *Salt as a Factor in the Confederacy*, 68, 125.

43. *O.R.*, series 4, vol. 1, 872–78.

44. *O.R.*, series 2, vol. 3, 178; *O.R.*, series 1, vol. 20, pt. 2, 478.

45. Maj. Gen. Simon B. Buckner in *O.R.*, series 1, vol. 32, pt. 2, 803.

46. Louis H. Manarin and Rufus J. Woolwine, "The Civil War Diary of Rufus J. Woolwine," *Virginia Magazine of History and Biography* 71, no. 4 (October 1963): 416–48; *O.R.*, series 1, vol. 20, pt. 1, 95–96.

47. *O.R.*, series 1, vol. 20, pt. 1, 104. On Carter's Raid, see William G. Piston, *Carter's Raid: An Episode of the Civil War in East Tennessee* (Johnson City, Tenn.: Overmountain Press, 1989).

48. *O.R.*, series 1, vol. 23, pt. 2, 853, quote on 906; *O.R.*, series 1, vol. 30, pt. 4, 676; Paludan, *Victims*, 85.

49. *O.R.*, series 1, vol. 23, pt. 2, 906.

50. Jones, *Rebel War Clerk's Diary*, 2:44, 73.

51. *O.R.*, series 1, vol. 31, pt. 3, 391.

52. *O.R.*, series 1, vol. 30, pt. 4, 617.

53. This was established by Special Orders No. 66, of the Adjutant and Inspector General's Office. *O.R.*, series 1, Vol. 33, 1233.

54. *O.R.*, series 1, vol. 32, pt. 2, 402.

55. Thomas D. Mays, *The Saltville Massacre* (Abilene, Tex.: WcWhiney Foundation Press, 1998), 47.

56. *O.R.*, series 1, vol. 32, pt. 2, 402; Chris J. Hartley, *Stoneman's Raid, 1865* (Winston Salem, N.C.: John F. Blair, 2010), 17.

57. *O.R.*, series 1, vol. 39, pt. 1, 553, 555–56.

58. On the conscription of black salt workers, see *O.R.*, series 1, vol. 51, pt. 2, 1047.

59. *O.R.*, series 1, vol. 39, pt. 1, 554; Mays, *Saltville Massacre*, 55, 62–63. According to Mays, Robertson transferred from Breckenridge's force to Georgia, where he would again be accused of executing Union prisoners.

60. "Rebel Barbarities—Colored Soldiers Murdered," *Liberator* 34, no. 47 (November 18, 1864): 188.

61. *O.R.*, series 1, vol. 49, pt. 1, 765. Ferguson would be captured and executed in 1865 for war crimes. On Ferguson, see Brian D. McKnight, *Confederate Outlaw: Champ Ferguson and the Civil War in Appalachia* (Baton Rouge: Louisiana State University Press, 2011).

62. Hartley, *Stoneman's Raid*, 9–20.

63. Lisa M. Brady, *War upon the Land: Military Strategy and the Transformation of Southern Landscapes during the American Civil War* (Athens: University of Georgia Press, 2012). See also Hurt, *Agriculture and the Confederacy*.

64. *O.R.*, series 1, vol. 34, pt. 1, 39; Mays, *Saltville Massacre*, 69–70; Hartley, *Stoneman's Raid*, 16–20.

65. *O.R.*, series 1, vol. 45, pt. 1, 808.

66. "Progress of the War," *New York Observer and Chronicle* 42, no. 51 (December 22, 1864): 406; "Weekly Summary: From the War," *Zion's Herald and Wesleyan Journal* 36, no. 1 (January 4, 1865): 3.

67. Nicholls, "Some Foundations of Economic Development," 291.

68. Jones, *Rebel War Clerk's Diary*, 2:367.

69. Mays, *Saltville Massacre*, 69–70.

70. Edward King, *The Great South* (1875; repr., Baton Rouge: Louisiana State University Press, 1972), 571; Andrew S. McCreath, *The Mineral Wealth of Virginia, Tributary to the Lines of the Shenandoah Valley and Norfolk and Western Railroad Companies* (Harrisburg, Penn.: Lane S. Hart, 1883), 13, 89–90.

71. Boyd, *Resources of South-West Virginia*, 104.

72. Nicholls, "Some Foundations of Economic Development," 291.

73. Brady, *War upon the Land*; Adam Wesley Dean, "Land-Development Politics and the American Civil War," chap. 3 in *An Agrarian Republic: Farming, Antislavery Politics, and Nature Parks in the Civil War Era* (Chapel Hill: University of North Carolina Press, 2015); Kathryn Shively Meier, *Nature's Civil War: Common Soldiers and the Environment in 1862 Virginia* (Chapel Hill: University of North Carolina Press, 2013); Jim Downs, *Sick from Freedom: African American Illness and Suffering during the Civil War and Reconstruction* (New York: Oxford University Press, 2012); Megan Kate Nelson, *Ruin Nation: Destruction and the American Civil War* (Athens: University of Georgia Press, 2012); Ted Steinberg, "The Great Food Fight," chap. 6 in *Down to Earth: Nature's Role in American History*, 3rd ed. (New York: Oxford University Press, 2012); Andrew McIlwaine Bell, *Mosquito Soldiers: Malaria, Yellow Fever, and the Course of the American Civil War* (Baton Rouge: Louisiana State University Press, 2010); and the collected essays in Brian Drake, ed., *The Blue, the Gray, and the Green: Toward an Environmental History of the Civil War* (Athens: University of Georgia Press, 2015).

74. Stephen Berry, "Forum: The Future of Civil War Era Studies," *Journal of the Civil War Era*, http://journalofthecivilwarera.org/forum-the-future-of-civil-war-era-studies.

75. Dan Kegley, "Memorial Service to Remember Troops of Oct. 2, 1864," *Smyth County News and Messenger* (Marion, Va.), September 30, 2009; Jannette Pippin, "Civil War Returns to Swansboro," *Daily News* (Jacksonville, N.C.), June 19, 2013; Rick Steelhammer, "Civil War Trail Winds Its Way to Kanawha Valley," *Charleston Gazette* (W.Va.), May 18, 2011; Jessica McCarthy, "Troops 'Battle' over Salt Again This Weekend," *News Herald* (Panama City, Fla.), Apr 22, 2012.

Chapter 5. Transportation: Roanoke, Railroads, and Appalachia on the Move

1. There is a vast literature on the expansion of American cities during the late nineteenth century. Some productive places to start include Sam Bass Warner, Jr., *The Urban Wilderness: A History of the American City* (Berkeley: University of California Press, 1995); Witold Rybczynski, *City Life* (New York: Scribner, 1996), esp. chap. 5; William Cronon,

Nature's Metropolis: Chicago and the Great West (New York: W. W. Norton, 1991); and Gunther Paul Barth, *City People: The Rise of Modern City Culture in Nineteenth Century America* (New York: Oxford University Press, 1980). For southern urban history, see David R. Goldfield, *Cotton Fields and Skyscrapers: Southern City and Region* (Baltimore: Johns Hopkins University Press, 1989).

2. William Goodell Frost, "Our Contemporary Ancestors in the Southern Mountains," *Atlantic Monthly* (March 1899): 311–19.

3. Starting places on this "discovery" era of Appalachian history include Allen Batteau, *The Invention of Appalachia* (Tucson: University of Arizona Press, 1990); Henry Shapiro, *Appalachia on Our Mind: The Southern Mountains and Mountaineers in the American Consciousness, 1870–1920* (Chapel Hill: University of North Carolina Press, 1978); and Cratis D. Williams, "The Southern Mountaineer in Fact and Fiction" (PhD diss., New York University, 1961).

4. Cronon, *Nature's Metropolis*.

5. Thomas Chapman, "Journal of a Journey through the United States, 1795–96," *Historical Magazine and Notes and Queries* 15 (June 1869): 357–68, in *Travels in the Old South: Selected from Periodicals of the Times*, ed. Eugene L. Schwaab and Jacqueline Bull (Lexington: University Press of Kentucky, 1973), 1:35.

6. Angus J. Johnston, II, "Virginia Railroads in April 1861," *Virginia Magazine of History and Biography* 22, no. 3 (August 1957): 312, 319; Kenneth W. Noe, *Southwest Virginia's Railroad: Modernization and the Sectional Crisis in the Civil War Era* (Tuscaloosa: University of Alabama Press, 2003); Raymond P. Barnes, *A History of the City of Roanoke* (Radford, Va.: Commonwealth Press, 1968), 4, 44.

7. Viator, "A Letter from a Virginian," *Southern Planter and Farmer* 40, no. 4 (April 1879): 184.

8. Talmage A. Stanley, *The Poco Field: An American Story of Place* (Urbana: University of Illinois Press, 2012), 24–27.

9. "The Mineral and Agricultural Wealth of Southwestern Virginia," *American Farmer* 1, no. 7 (January 1867): 212. See also "The Mineral and Agricultural Wealth of Southwestern Virginia," *Southern Planter and Farmer* 1, no. 2 (March 1867): 109.

10. Edward King, *The Great South: A Record of Journeys in Louisiana, Texas, the Indian Territory, Missouri, Arkansas, Mississippi, Alabama, Georgia, Florida, South Carolina, North Carolina, Kentucky, Tennessee, Virginia, West Virginia, and Maryland* (Hartford, Conn.: American Publishing, 1875), 578.

11. Charles Rufus Boyd, "The Mineral Wealth of Southwestern Virginia," *Transactions of the Society of Mining Engineers of the American Institute of Mining, Metallurgical and Petroleum Engineers* (1877): 81–92; Charles Rufus Boyd, *Resources of South-West Virginia* (New York: John Wiley & Sons, 1881).

12. King, *Great South*, 569–78, quote on 576; Ronald D. Eller, *Miners, Millhands, and Mountaineers: Industrialization of the Appalachian South, 1880–1930* (Knoxville: University of Tennessee Press, 1982), 69–70.

13. "Roanoke: The New Virginia City," *Roanoke Leader*, September 7, 1882.

14. "The Coal and Iron Interest of Southwest Virginia," *Roanoke Leader*, September 21, 1882.

15. On the long-term use of local coal in local locomotives, see Robert C. Post, "'The Last Steam Railroad in America': Shaffers Crossing, Roanoke, Virginia, 1958," *Technology and Culture* 44, no. 3 (July 2003): 560–65.

16. Randal L. Hall, *Mountains on the Market: Industry, the Environment, and the South* (Lexington: University Press of Kentucky, 2012), 86–88, 111.

17. David C. Hsiung, *Two Worlds in the Tennessee Mountains: Exploring the Origins of Appalachian Stereotypes* (Lexington: University Press of Kentucky, 1997), 183.

18. John Alexander Williams, *Appalachia: A History* (Chapel Hill: University of North Carolina Press, 2002), 232; Crandall A. Shifflett, *Coal Towns: Life, Work, and Culture in Company Towns of Southern Appalachia, 1880–1960* (Knoxville: University of Tennessee Press, 1991), 33–66.

19. Stuart Seely Sprague, "The Great Appalachian Iron and Coal Town Boom of 1889–1893," *Appalachian Journal* 4, nos. 3–4 (Spring/Summer 1977): 220–21; Harry Caudill, *Night Comes to the Cumberlands: A Biography of a Depressed Area* (Ashland, Ky.: Jesse Stuart Foundation, 2001), 77–78.

20. George O. Robinson, Jr., *The Oak Ridge Story: The Saga of a People Who Share in History* (Kingsport, Tenn.: Southern Publishers, 1950), 39–41.

21. Tom Lee, *The Tennessee-Virginia Tri-Cities: Urbanization in Appalachia, 1900–1950* (Knoxville: University of Tennessee Press, 2005).

22. *New York Herald* quoted in ibid., 216.

23. Howard N. Rabinowitz, "Continuity and Change: Southern Urban Development, 1860–1900," in *The City in Southern History: The Growth of Urban Civilization in the South*, ed. Blaine A. Brownell and David R. Goldfield (Port Washington, N.Y.: Kennikat Press, 1977): 93, table 4-1; Eller, *Miners, Millhands, and Mountaineers*, 69–70.

24. E. G. D., "Roanoke City Crowded, by Delegates to the Democratic Convention," *New York Times*, August 4, 1887.

25. Warner in *Hartford Courant*, quoted in Rand Dotson, "New South Boomtown: Roanoke, Virginia, 1882-1884," *Virginia Magazine of History and Biography* 116, no. 2 (2008): 164.

26. Philip J. Dreyfus, "Urban Genesis," chap. 2 in *Our Better Nature: Environment and the Making of San Francisco* (Norman: University of Oklahoma Press, 2008); Thomas G. Andrews, "The Reek of the New Industrialism," chap. 2 in *Killing for Coal: America's Deadliest Labor War* (Cambridge: Harvard University Press, 2008); Andrew C. Isenberg, "Banking on Sacramento: Urban Development, Flood Control, and Political Legitimization," chap. 2 in *Mining California: An Ecological History* (New York: Hill and Wang, 2005); Patricia Nelson Limerick, *The Legacy of Conquest: The Unbroken Past of the American West* (New York: W. W. Norton, 1987), 71–73.

27. Cronon, *Nature's Metropolis*.

28. Barnes, *History of the City of Roanoke*, 104.

29. "Roanoke: The New Virginia City," and advertisements throughout *Roanoke Leader*, September 7, 1882.

30. Scott Reynolds Nelson, *Steel Drivin' Man: John Henry, the Untold Story of an American Legend* (New York: Oxford University Press, 2006), 26, 70, 98.

31. Barnes, *History of the City of Roanoke*, 132.

32. Williams, *Appalachia*, 237–38.

33. Clare White, *Roanoke, 1740–1982* (Roanoke, Va.: Roanoke Valley Historical Society, 1982), 70–71.

34. "Roanoke Machine Works," *Roanoke Leader*, August 30, 1883.

35. W. E. Christian, "Salem: The Growing Town of Southwest Virginia," *Forum* (New York) 9, no. 6 (August 1890): 28, quote on 25.

36. Rabinowitz, "Continuity and Change," 96.

37. Steven E. Nash, *Reconstruction's Ragged Edge: The Politics of Postwar Life in the Southern Mountains* (Chapel Hill: University of North Carolina Press, 2016), 151–58.

38. "Railroads," *Roanoke Leader*, February 14, 1884.

39. Stanley, *Poco Field*, 24–28.

40. Quoted in Andrew S. McCreath, *The Mineral Wealth of Virginia Tributary to the Lines of the Norfolk and Western and Shenandoah Valley Railroad Companies* (Harrisburg, Penn.: Lane S. Hart, 1884), iii.

41. Ibid., 105.

42. Andrew S. McCreath and E. V. d'Invilliers, *The New River–Cripple Creek Mineral Region of Virginia* (Harrisburg, Penn.: Harrisburg Publishing, 1887), 163.

43. Williams, *Appalachia*, 232–33; Eller, *Miners, Millhands, and Mountaineers*, 69–70, 73–74; county population figures compiled from the University of Virginia Library's Historical Census Browser, http://mapserver.lib.virginia.edu/, last accessed August 13, 2016. Similar explosive population growth followed the extension of other rail lines into Appalachian coal belts. For Kentucky examples in later decades, see Caudill, *Night Comes to the Cumberlands*, 106.

44. Chad Montrie, *Making a Living: Work and Environment in the United States* (Chapel Hill: University of North Carolina Press, 2008), 75.

45. Hall, *Mountains on the Market*, 90.

46. Kate Brown, *Dispatches from Dystopia: Histories of Places Not Yet Forgotten* (Chicago: University of Chicago Press, 2015), 105.

47. "Roanoke Coal Fields Open," *New York Times*, July 10, 1902.

48. Hall, *Mountains on the Market*, 105–6.

49. "Iron Making in Virginia," *Roanoke Leader*, June 28, 1883; quote in "What a First-Class Basic Steel Works Would Do for Virginia," *Roanoke Leader*, September 14, 1882.

50. "The Big Gun Foundry," "The Cotton Factory," and "State Normal School," *Roanoke Leader*, February 21, 1884.

51. "Underworking and Underselling the North," *Roanoke Leader*, February 21, 1884.

52. On major metropolitan centers struggling with these issues, see Catherine McNeur, *Taming Manhattan: Environmental Battles in the Antebellum City* (Cambridge: Harvard University Press, 2014); Martin V. Melosi, *The Sanitary City: Urban Infrastructure in America from Colonial Times to the Present* (Baltimore: Johns Hopkins University Press, 2000), esp. chaps. 4–9; and Joel A. Tarr, *The Search for the Ultimate Sink: Urban Pollution in Historical Perspective* (Akron, Ohio: University of Akron Press, 1996).

53. White, *Roanoke, 1740–1982*, 71–72.

54. "Sanitary," *Roanoke Leader*, September 21, 1882.

55. White, *Roanoke, 1740–1982*, 70–71, 87; Dotson, "New South Boomtown," 152, 175–77; Barnes, *History of the City of Roanoke*, 109–11, quote on 116.

56. Barnes, *History of the City of Roanoke*, 118.

57. "Severe Cloudburst at Roanoke," *New York Times*, August 24, 1892.

58. Barnes, *History of the City of Roanoke*, 186.

59. Grouse, "Quail in Roanoke and Wythe," *Forest and Stream* 33, no. 2 (January 16, 1890): 513.

60. Dotson, "New South Boomtown," 167–69.

61. Ann Field Alexander, "'Like an Evil Wind': The Roanoke Riot of 1893 and the Lynching of Thomas Smith," *Virginia Magazine of History and Biography* 100, no. 2 (April 1992): 178–79.

62. Grouse, "Quail in Roanoke and Wythe," 513.

63. David R. Goldfield, *Cotton Fields and Skyscrapers: Southern City and Region* (Baltimore: Johns Hopkins University Press, 1989), 4.

64. Jeff Biggers, *The United States of Appalachia: How Southern Mountaineers Brought Independence, Culture, and Enlightenment to America* (New York: Shoemaker and Hoard, 2006), 147–56, 162–64. On the growth of the southern textile industry and life in mill towns, see Allen Tullos, *Habits of Industry: White Culture and the Transformation of the Carolina Piedmont* (Chapel Hill: University of North Carolina Press, 1989); Jacquelyn Dowd Hall, James Leloudis, Robert Korstad, Mary Murphy, Lu Ann Jones, and Christopher B. Daly, *Like a Family: The Making of a Southern Cotton Mill World* (Chapel Hill: University of North Carolina Press, 1987).

65. Alexander, "Like an Evil Wind," 180–81; Barnes, *History of the City of Roanoke*, 246–47.

66. Alexander, "Like an Evil Wind," 173–206; Suzanne Lebsock, *A Murder in Virginia: Southern Justice on Trial* (New York: W. W. Norton, 2003), 62–64. For the continuation of Jim Crow racial strictures and the concentrated prejudices of the countryside in early twentieth-century Roanoke, see Beth Macy, *Truevine: Two Brothers, a Kidnapping, and a Mother's Quest: A True Story of the Jim Crow South* (New York: Little, Brown, 2016), esp. 43–44.

67. "The Roanoke Riot Cases," *New York Times*, Nov. 23, 1893, 5; Hall, *Mountains on the Market*, 100–101.

68. White, *Roanoke, 1740–1982*, 75–76, 84.

69. Ibid., 84–86, 101; Barnes, *History of the City of Roanoke*, 167, 171; Dotson, "New South Boomtown," 182; "Roanoke College," *Southern Planter* 52, no. 7 (July 1891): 391.

70. Lon Savage, *Thunder in the Mountains: The West Virginia Mine War, 1920–21* (Pittsburgh: University of Pittsburgh Press, 1990), 142–44; White, *Roanoke, 1740–1982*, 66–67.

71. For general background of the Appalachian Regional Commission, including county lists and maps, see its website at www.arc.gov/.

72. Williams, *Appalachia*, 341.

73. William A. Link, *A Hard Country and a Lonely Place: Schooling, Society, and Reform in Rural Virginia, 1870–1920* (Chapel Hill: University of North Carolina Press, 1986), 74; White, *Roanoke, 1740–1982*, 84–86, 101; Dotson, "New South Boomtown," 182.

74. For a statistical analysis of urban Appalachia and an argument for the importance of towns in regional history, see Ann DeWitt Watts, "Cities and Their Place in Southern Appalachia," *Appalachian Journal* 8, no. 2 (Winter 1981): 105–18.

75. Steven Conn, *Americans against the City: Anti-Urbanism in the Twentieth Century* (New York: Oxford University Press, 2014), 1–2.

Chapter 6. Scenery: Recreation and Tourism on Grandfather Mountain

1. Shepherd M. Dugger, *The War Trails of the Blue Ridge, Containing an Authentic Description of the Battle of Kings Mountain, the Incidents Leading up to and the Echoes of the Aftermath of This Epochal Engagement, and Other Stories Whose Scenes Are Laid in the Blue Ridge* (Banner Elk, N.C.: Shepherd M. Dugger, 1932), 147. Beginning in the 1870s, northern tourism in the southern Appalachians grew rapidly, spurred by increasing railroad connections and people's growing desire to escape urban industrial settings. Northwestern North Carolina was particularly popular with tourists from Philadelphia. See

Kevin O'Donnell and Helen Hollingsworth, eds., *Seekers of Scenery: Travel Writing from Southern Appalachia, 1840–1900* (Knoxville: University of Tennessee Press, 2004), 20–21, 194–96, 307–17.

2. Dugger, *War Trails of the Blue Ridge*, 147–48; Howard E. Covington Jr., *Linville: A Mountain Home for 100 Years* (Linville, N.C.: Linville Resorts, 1992), 21.

3. For an overview of this tourism in western North Carolina, see Richard D. Starnes, *Creating the Land of the Sky: Tourism and Society in Western North Carolina* (Tuscaloosa: University of Alabama Press, 2005).

4. Roderick Frazier Nash, *Wilderness and the American Mind*, 4th ed. (New Haven, Conn.: Yale University Press, 2001), esp. chaps. 3–7; William Cronon, "The Trouble with Wilderness; or, Getting Back to the Wrong Nature," in *Uncommon Ground: Rethinking the Human Place in Nature* (New York: W. W. Norton, 1996), 72–76.

5. Gregg Mitman, "Hay Fever Holiday" and "The Last Resorts," chaps. 1 and 3 in *Breathing Space: How Allergies Shape Our Lives and Landscapes* (New Haven, Conn.: Yale University Press, 2007); Steven Conn, *Americans against the City: Anti-Urbanism in the Twentieth Century* (New York: Oxford University Press, 2014), 139–44.

6. For example, see Margaret Brown, *The Wild East: A Biography of the Great Smoky Mountains* (Gainesville: University Press of Florida, 2001); Anne Mitchell Whisnant, *Super-Scenic Motorway: A Blue Ridge Parkway History* (Chapel Hill: University of North Carolina Press, 2005); Timothy Silver, *Mount Mitchell and the Black Mountains: An Environmental History of the Highest Peaks in Eastern America* (Chapel Hill: University of North Carolina Press, 2003); Sue Eisenfeld, *Shenandoah: A Story of Conservation and Betrayal* (Lincoln: University of Nebraska Press, 2015); Sara M. Gregg, *Managing the Mountains: Land Use Planning, the New Deal, and the Creation of a Federal Landscape in Appalachia* (New Haven, Conn.: Yale University Press, 2010); Daniel Pierce, *The Great Smokies: From Natural Habitat to National Park* (Knoxville: University of Tennessee Press, 2000). For studies of private development of mountain tourist attractions, see Starnes, *Creating the Land of the Sky*; Christopher Eklund, "Private Paths to Protecting Places: The Creation of a Conservation Infrastructure in the American South since 1889," PhD diss. (Auburn University, 2015).

7. Linville Improvement Company, *Linville* (Linville, N.C.: Linville Improvement Company, 1888), 5, from University Library, University of North Carolina at Chapel Hill, https://archive.org/details/linvilleoolinv.

8. André Michaux, "Journal of André Michaux, 1787–1796," ed. Charles S. Sargent, *Proceedings of the American Philosophical Society* 26 (1888): 112; François André Michaux, *Travels to the Westward of the Allegany Mountains, in the States of Ohio, Kentucky, and Tennessee, in the Year 1802* (London: Barnard & Sultzer, 1805), 95; Joseph Ewan and Nesta Ewan, "John Lyon, Nurseryman and Plant Hunter, and His Journal, 1799–1814," *Transactions of the American Philosophical Society* 53 (1963): 8; Elisha Mitchell, *Diary of a Geological Tour by Dr. Elisha Mitchell in 1827 and 1828*, ed. Kemp P. Battle, James Sprunt Historical Monograph 6 (Chapel Hill, N.C.: University Press, 1905), 35; Ronald H. Petersen, "Moses Ashley Curtis's 1839 Expedition into the North Carolina Mountains," *Castanea* 53 (June 1988): 114; Asa Gray, "Notes of a Botanical Excursion to the Mountains of North Carolina, &c.; with Some Remarks on the Botany of the Higher Alleghany Mountains," *American Journal of Science and Arts* 42 (April 1842): 31, 36; Winona H. Welch, "The Moss Foray in North Carolina, June 13–15, 1936," *Bryologist* 39 (November/December 1936): 122; A. A. Heller, "Notes on the Flora of North Carolina," *Bulletin of the Torrey Botanical Club* 18

(June 1891): 190–91; J. H. Redfield, "Notes of a Botanical Excursion into North Carolina," *Bulletin of the Torrey Botanical Club* 6 (July/August 1879): 331–39; John W. Harshberger, "An Ecological Study of the Flora of Mountainous North Carolina," *Botanical Gazette* 36 (November 1903): 379.

9. University of North Carolina Alumni Questionnaire, 1923, NC Box File—Shepherd Dugger #1, Stirling Collection, James H. Carson Library, Lees-McRae College, Banner Elk, North Carolina (hereafter SCLMC); Shepherd Dugger, "Autobiography," unpublished typescript, n.d., 1, 8, 10, NC Box File—Shepherd Dugger #1, SCLMC; Leslie Banner Cottingham and Carol Lowe Timblin, *The Bard of Ottaray* (Banner Elk, N.C.: Puddingstone Press, 1979), 12, 42; Sanna R. Gaffney, ed., *The Heritage of Watauga County North Carolina* (Winston-Salem, N.C.: Hunter Publishing, 1984), 101. With typical Dugger humor, on his UNC alumni questionnaire the author described his wife Margaret as educated "in the woods," at "Cornfield College," with a "'Tater'-patch degree." The Dugger manuscript is a brief document outlining his career as a hotel operator, tourism promoter, and after 1900 as an official in the newly formed Avery County government.

10. Dugger, "Autobiography," 9.

11. Brochure in Waightstill Avery, et al., *Avery County Heritage: Biographies and Genealogies* (Banner Elk, N.C.: Puddingstone Press, 1976), 70.

12. Dugger, "Autobiography," 8.

13. Anonymous, "A Sunset on Grandfather Mountain," *Christian Observer* 84, no. 50 (December 9, 1896): 21.

14. Shepherd Dugger, *The Balsam Groves of the Grandfather Mountain: A Tale of the Western North Carolina Mountains, Together with Information Relating to the Section and Its Hotels, also a Vocabulary of Indian Names and a List of Altitudes of Important Mountains, etc.* (Philadelphia: John C. Winston, 1907), 62–63. The novel went through several editions, and the 1907 version was the most expansive (as suggested by its long title).

15. Ibid., 115.

16. Ibid., 233–34.

17. Edward W. Bok, "The Pen of a Mountaineer," *Ladies' Home Journal* 11, no. 8 (July 1894): 12.

18. Anonymous, "Sunset on Grandfather Mountain," 20.

19. For an analysis of early and mid-nineteenth-century beliefs about the health of southern landscapes, see Conevery Bolton Valenčius, *The Health of the Country: How American Settlers Understood Themselves and Their Land* (New York: Basic Books, 2002).

20. Albert Way, *Conserving Southern Longleaf: Herbert Stoddard and the Rise of Ecological Land Management* (Athens: University of Georgia Press, 2011), 20–31; Gregg Mitman, "Hay Fever Holiday: Health, Leisure, and Place in Gilded-Age America," *Bulletin of the History of Medicine* 77 (2003): 600–635; Mitman, "Hay Fever Holiday" and "The Last Resorts"; Linda Nash, "Finishing Nature: Harmonizing Bodies and Environments in Late-Nineteenth Century California," *Environmental History* 8, no. 1 (2003): 25–52; Linda Nash, *Inescapable Ecologies: A History of Environment, Disease, and Knowledge* (Berkeley: University of California Press, 2006), 56–57. The 1880s and 1890s were also the peak of health tourism at the famous Arkansas Hot Springs according to Richard Sellars, *Preserving Nature in the National Parks: A History* (New Haven: Yale University Press, 1997), 18.

21. John Alexander Williams, *West Virginia: A History* (New York: W. W. Norton, 1984), 157–58; Jeanette Keith, "Good Times: Vacationing at Red Boiling Springs," in *Rural Life*

and Culture in the Upper Cumberland, ed. Michael E. Birdwell and W. Calvin Dickinson (Lexington: University Press of Kentucky, 2004): 178–95.

22. Starnes, *Creating the Land of the Sky*, 57; S. T. Kelsey and C. C. Hutchinson, *The Blue Ridge Highlands of Western North Carolina* (Atlanta: J. P. Harrison, 1878).

23. Edgar Tufts, "The Lees-McRae Institute at Banner Elk, N.C.," *Christian Observer* 92, no. 1 (January 6, 1904): 11.

24. Edgar Tufts, "The Lees-McRae Institute," *Christian Observer* 92, no. 40 (October 5, 1904): 8.

25. For prominent examples, see Ronald D. Eller, *Miners, Millhands, and Mountaineers: Industrialization of the Appalachian South, 1880–1930* (Knoxville: University of Tennessee Press, 1982); Paul Salstrom, *Appalachia's Path to Dependency: Rethinking a Region's Economic History, 1730–1940* (Lexington: University Press of Kentucky, 1994); Ronald Lewis, *Transforming the Appalachian Countryside: Railroads, Deforestation, and Social Change in West Virginia, 1880–1920* (Chapel Hill: University of North Carolina Press, 1998); and Starnes, *Creating the Land of the Sky*, 9–91. For an overview of the degree of ecological interconnectedness in Appalachia, see Donald E. Davis, *Where There Are Mountains: An Environmental History of the Southern Appalachians* (Athens: University of Georgia Press, 2000). An especially good introduction to the diverse body of scholarship on the growing commercial ties of southern Appalachia at the end of the nineteenth century is a collection of essays edited by Mary Beth Pudup, Dwight Billings, and Altina Waller: *Appalachia in the Making: The Mountain South in the Nineteenth Century* (Chapel Hill: University of North Carolina Press, 1995).

26. David Hsiung, *Two Worlds in the Tennessee Mountains: Exploring the Origins of Appalachian Stereotypes* (Lexington: University Press of Kentucky, 1997); Lewis, *Transforming the Appalachian Countryside*.

27. Statistical profile assembled from the Historical Census Browser, UVA Library, http://mapserver.lib.virginia.edu/. Statewide, the average farm was 127 acres, while in Caldwell, Mitchell, and Watauga it was 110 acres. There were 1.89 acres of unimproved land for each improved acre across North Carolina; in the three counties the ratio was 2.15 to one.

28. John L. Hartley, *Walking for Health and Traveling to Eternity, Combined with Singing on the Mountain* (Linville, N.C., n.d.), 3–5, NC Box File—Grandfather Mountain, SCLMC; Avery County Historical Society, *Avery County Heritage: Biographies, Genealogies and Church Histories* (Banner Elk, N.C.: Puddingstone Press, 1981), 137; Horton Cooper, *History of Avery County North Carolina* (Asheville, N.C.: Biltmore Press, 1964), 8.

29. Hartley, *Walking for Health and Traveling to Eternity*, 2; Cooper, *History of Avery County*, 32.

30. The best study of this epoch and the timber industry's environmental and social impact in Appalachia is Lewis, *Transforming the Appalachian Countryside*. For western North Carolina, see Kathryn Newfont, *Blue Ridge Commons: Environmental Activism and Forest History in Western North Carolina* (Athens: University of Georgia Press, 2012), 42–48.

31. Michael Williams, *Americans and Their Forests: A Historical Geography* (Cambridge: Cambridge University Press, 1989), 239.

32. Margaret Brown, *Wild East*, 55.

33. Timothy Silver, *Mount Mitchell and the Black Mountains: An Environmental History of the Highest Peaks in Eastern America* (Chapel Hill: University of North Carolina Press, 2003), 136–43.

34. U.S. Department of Agriculture, *A Message from the President of the United States Transmitting a Report of the Secretary of Agriculture in Relation to the Forests, Rivers, and Mountains of the Southern Appalachian Region* (Washington D.C.: GPO, 1902), 24.

35. Margaret Morley, *The Carolina Mountains* (Boston: Houghton Mifflin, 1913), 24.

36. Ibid., 370–78.

37. U.S. Department of Agriculture, *Message from the President*, 50.

38. Ibid., plate 6, pts. 1 and 2, facing p. 19; plate 7, facing p. 20.

39. Ibid., 49.

40. Eller, *Miners, Millhands, and Mountaineers*, 48; Loren M. Wood, *Beautiful Land of the Sky: John Muir's Forgotten Eastern Counterpart, Harlan P. Kelsey* (Bloomington, Ind.: iUniverse, 2013), 62–63, 65–66.

41. Covington, *Linville*, 3–7; Logan H. Brown, "An Historical Examination of Tourism Marketing Imagery in Western North Carolina and Its Impacts on Cultural Interpretation of the Landscape (master's thesis, Appalachian State University, 2001), 39; Steven E. Nash, *Reconstruction's Ragged Edge: The Politics of Postwar Life in the Southern Mountains* (Chapel Hill: University of North Carolina Press, 2016), 37, 164–67; Whisnant, *Super-Scenic Motorway*, 271. Historian Richard Starnes has suggested that Highlands and Linville (along with attractions like Asheville's Grove Park Inn and the Cloudland Hotel on Roan Mountain) were products of a region-wide attempt to define western North Carolina as "the land of the sky," a tourist mecca that blended the natural with the improved. Starnes, *Creating the Land of the Sky*, 35–63.

42. Wood, *Beautiful Land of the Sky*, 479.

43. Herbert Francis Sherwood, "Directing the Great Human River," *Outlook* 96, no. 15 (December 10, 1910): 824–25; Hugh MacRae, *Vitalizing the Nation and Conserving Human Units through the Development of Agricultural Communities* (Philadelphia: American Academy of Political and Social Science, 1916).

44. *Eseeola Inn and Annex* (Philadelphia: Loughead, n.d.), 11, from State Library of North Carolina, https://archive.org/details/eseeolainnannexlooesee; Fred A. Olds, "Tramping in North Carolina Mountains," *Forest and Stream* 81, no. 24 (December 13, 1913): 747.

45. "Pamphlets and Books Received," *Christian Union* 45, no. 21 (May 21, 1892): 1000; "Pamphlets and Books Received," *Christian Union* 47, no. 21 (May 27, 1893): 1038.

46. Quoted in Covington, *Linville*, 10.

47. Logan H. Brown, "Historical Examination of Tourism Marketing," fig. 11, 136.

48. Dugger, *War Trails of the Blue Ridge*, 147. While Dugger's statement appears derogatory, he was an advocate of Linville's growth, as the rail connection and better roads created for the town improved his own hotel's business by bringing in greater numbers of tourists. To this end, in *The Balsam Groves of the Grandfather Mountain* Dugger touted Linville as wonderful vacation spot and reminded readers who appreciated solitude that his own more secluded hotel was just a short distance to the north.

49. Covington, *Linville*, 17; Logan H. Brown, "Historical Examination of Tourism Marketing," 41. Rittenhouse, a resident of Cairo, Illinois, had never visited Linville and created her descriptions of Appalachia from Linville Resorts pamphlets and her imagination. "In the Afterglow" apparently never found a publisher, though Rittenhouse did publish a short story collection (*A Candid Critic, and Other Stories for Girls*) in 1897. According to Rittenhouse's biographer, Dugger did not even take second place: "A Miss Jones

of Berlin, Germany, won the 2nd prize of 250.00." See Richard Lee Strout, ed., *Maud* (New York: Macmillan, 1939), 534–35, 560–61; Maude [*sic*] Rittenhouse Mayne, *A Candid Critic, and Other Stories for Girls* (Philadelphia: American Baptist Publication Society, 1897).

50. Starnes, *Creating the Land of the Sky*, 37–42.

51. Though all editions of *Balsam Groves* list "Banner Elk: Shepherd M. Dugger," on the title page, the J. B. Lippincott Company of Philadelphia printed the 1892 and 1895 editions, John C. Winston of the same city printed the 1907 edition, and the Observer Printing House in Charlotte released the final edition. See Leslie Cottingham and Carol Timblin, *The Bard of Ottaray: The Life, Letters and Documents of Shepherd Monroe Dugger* (Banner Elk, N.C.: Puddingstone Press, 1979), 16–17.

52. Dugger, *Balsam Groves*, 85–86.

53. Ibid., 116–17, 137.

54. Cratis D. Williams, "The Southern Mountaineer in Fact and Fiction" (PhD diss., New York University, 1961); Henry D. Shapiro, *Appalachia on Our Mind: The Southern Mountains and Mountaineers in the American Consciousness, 1870–1920* (Chapel Hill: University of North Carolina Press, 1986); Ronald L. Lewis, "Beyond Isolation and Homogeneity: Diversity and the History of Appalachia," in *Confronting Appalachian Stereotypes: Back Talk from an American Region*, ed. Dwight Billings, Gurney Norman, and Katherine Ledford (Lexington: University Press of Kentucky, 1999), 21–22; John Alexander Williams, *Appalachia: A History* (Chapel Hill: University of North Carolina Press, 2002), 197–202; Allen Batteau, *The Invention of Appalachia* (Tucson: University of Arizona Press, 1990), 81–82.

55. *Eseeola Inn, 1891* (Linville, N.C.: Linville Improvement Company, 1892), 3, in Linville Improvement Company papers (hereafter referred to as LIC). Subsequent to the author's examination, many of these papers, once in the possession of Grandfather Mountain Inc. in Linville, have been donated to the Southern Historical Collection at the University of North Carolina at Chapel Hill.

56. In this manner, they are perhaps more similar to the mountaineers described in Horace Kephart, *Our Southern Highlanders* (New York: Outing Publishing, 1913); Emma Bell Miles, *The Spirit of the Mountains* (1905; repr., Knoxville: University of Tennessee Press, 1975); and John C. Campbell, *The Southern Highlander and His Homeland* (New York: Russell Sage Foundation, 1921). For a connection of these three works, see Ted Olson, "Literature," in *High Mountains Rising: Appalachia in Time and Place*, ed. Richard A. Straw and Tyler H. Blethen (Urbana: University of Illinois Press, 2004), 170.

57. Mitman, *Breathing Space*.

58. Dugger, *Balsam Groves of the Grandfather Mountain*, 223–25.

59. Hartley, *Walking for Health and Traveling to Eternity*, 21.

60. G. C. Connor, "Trout in Western North Carolina," *Forest and Stream* 26, no. 25 (July 15, 1886): 489.

61. Covington, *Linville*, 34–39.

62. *Eseeola Inn, 1891*, 3, LIC; *The Eseeola Inn, Linville, North Carolina* (Bristol, Tenn.: King Printing, 1908), 3, LIC.

63. Melville Anderson, "The Conversation of John Muir," *American Museum Journal* 15 (1915): 119.

64. *Eseeola Inn and Annex*, 9.

65. Linville Improvement Company, *Linville*, 9, 11.

66. Covington, *Linville*, 15–16; Logan H. Brown, "Historical Examination of Tourism Marketing," 40; Barry M. Buxton, *A Village Tapestry: The History of Blowing Rock* (Boone, N.C.: Appalachian Consortium Press, 1989), 7.

67. James I. Vance, "The Plum Tree Conference, in the Mountains of North Carolina," *Christian Observer* 92, no. 36 (September 7, 1904): 11; Frank W. Bicknell, "New Forest Roads," *Forest and Stream* 75, no. 5 (July 30, 1910): 174.

68. S. T. Kelsey and C. C. Hutchinson, *The Blue Ridge Highlands of Western North Carolina* (Atlanta: J. P. Harrison, 1878), 1; Richard D. Starnes, "Sanitarium, Railroads, and the New South," chap. 1 in *Creating the Land of the Sky*.

69. Linville Improvement Company, *Linville*, 21; *Eseeola Inn and Annex*, 13.

70. Julian Scheer and Elizabeth Black, *Tweetsie: The Blue Ridge Stemwinder* (Charlotte, N.C.: McNally and Loftin, 1958), 8–19; Mallory Hope Ferrell, *Tweetsie Country: The East Tennessee and Western North Carolina Railroad* (Johnson City, Tenn.: Overmountain Press, 1979), 48–49.

71. Covington, *Linville*, 61–62; quote in "May Make a Park of Grandfather Mountain," *Cleveland Star* (Shelby, N.C.), April 9, 1923, Vertical Clippings File—Grandfather Mountain, W. L. Eury Appalachian Collection, Carol Grotnes Belk Library, Appalachian State University, Boone, North Carolina (hereafter ASU).

72. Gaffney, *Heritage of Watauga County*, 71–72; Ferrell, *Tweetsie Country*, 30–31.

73. Stockholder Meeting Minutes, Linville Improvement Company, February 9, 1911 through April 14, 1952, 30, LIC.

74. Johnny Cooper, interviewed by Randy Johnson, 1990, typescript, Vertical File—Interviews, Grandfather Mountain Stewardship Foundation files, Linville, North Carolina.

75. Ibid.; Buxton, *Village Tapestry*, 34.

76. Ted Shook, "Grandfather Mtn. Acquired Its First Haircut in 1930," *Watauga Democrat*, January 29, 1976, Vertical File—Grandfather Mountain, ASU.

77. Harlan P. Kelsey, "What Can Be Done," *Chautauquan* 51, no. 1 (June 1908): 55–57; Wood, *Beautiful Land of the Sky*, 69, 77–99.

78. E. P. Powell, "Our Native Plants," *Independent . . . Devoted to the Consideration of Politics, Social and Economic Tendencies, History, Literature, and the Arts* 42, no. 2169 (June 26, 1890): 31; advertisement, *Southern Planter* 51, no. 12 (December 1, 1890): 586; Wood, *Beautiful Land of the Sky*, 85–89.

79. "May Make a Park of Grandfather Mountain"; Wood, *Beautiful Land of the Sky*, 96, 105, 229–37; Michael Frome, *Strangers in High Places: The Story of the Great Smoky Mountains* (Garden City, N.Y.: Doubleday, 1966), 181, 213; Sarah Mittlefehldt, *Tangled Roots: The Appalachian Trail and American Environmental Politics* (Seattle: University of Washington Press, 2013), 48. Samuel Kelsey was the first southern member of the Appalachian Mountain Club.

80. Sue Eisenfeld, *Shenandoah: A Story of Conservation and Betrayal* (Lincoln: University of Nebraska Press, 2015); Katrina Powell, ed., *"Answer at Once": Letters of Mountain Families in Shenandoah National Park, 1934–1938* (Charlottesville: University of Virginia Press, 2013); Margaret Brown, *Wild East*; Frome, *Strangers in High Places*.

81. Frome, *Strangers in High Places*, 212–13; Wood, *Beautiful Land of the Sky*, 477–516.

82. Stockholder Meeting Minutes, Linville Improvement Company, 30–60.

83. "Grandfather Mountain: Monarch in a Land of Majestic Peaks!," *Touring* (1935): 14–15.

84. Paul S. Sutter, *Driven Wild: How the Fight against Automobiles Launched the Modern Wilderness Movement* (Seattle: University of Washington Press, 2002), 37–38; Gregg Mitman, *Reel Nature: America's Romance with Wildlife on Film* (Cambridge: Harvard University Press, 1999), 95; Sellars, *Preserving Nature in the National Parks*, 52–53.

85. Covington, *Linville*, 78–79.

86. Batteau, *Invention of Appalachia*, 87; Sutter, *Driven Wild*, esp. chapters 1 and 2.

87. Hight C. Moore, "Grandfather Mount; A Personality Peak," *Charlotte Observer*, August 27, 1939, Vertical File—Grandfather Mountain, ASU.

88. Ruth Moore, "Lofty Mountain Thrills Traveler," *Greensboro Daily News*, August 29, 1937, NC Box File—Grandfather Mountain, SCLMC.

89. Stockholder Meeting Minutes, Linville Improvement Company, 156–57.

90. Hartley, *Walking for Health and Traveling to Eternity*, 3, 6.

91. Roderick Frazier Nash, *Wilderness and the American Mind*, 141–60, 200–209; Sutter, *Driven Wild*, esp. chapter 2.

92. Celeste Ray, *Highland Heritage: Scottish Americans in the American South* (Chapel Hill: University of North Carolina Press, 2001), 59–60, 150–51.

93. Anne Mitchell Whisnant, "Public and Private Tourism Development in 1930s Appalachia: The Blue Ridge Parkway Meets Little Switzerland," in *Southern Journeys: Tourism, History, and Culture in the Modern South*, ed. Richard D. Starnes (Tuscaloosa: University of Alabama Press, 2003), 90–95; Carl A. Schenck, *Birth of Forestry in America: Biltmore Forest School, 1899–1913* (Durham, N.C.: Forest History Society, 1974); Howard E. Covington Jr., *Lady on the Hill: How Biltmore Estate Became an American Icon* (New York: Wiley, 2006). For a description of Chimney Rock State Park, see its website: www.chimneyrockpark.com/.

94. On state acquisition of Chimney Rock, see "Chimney Rock State Park History," North Carolina State Parks, www.ncparks.gov/chimney-rock-state-park/history.

95. Mittlefehldt, *Tangled Roots*; Conn, *Americans against the City*, 139–44; Paul Sutter, "Wilderness as Regional Plan: Benton MacKaye," chap. 5 in *Driven Wild*.

96. "What Is Grandfather Mountain?," www.grandfather.com/about-grandfather -mountain/what-is-grandfather-mountain.

Chapter 7. Tobacco: Making Ground for an International Crop

1. Ann K. Ferrell, *Burley: Kentucky Tobacco in a New Century* (Lexington: University Press of Kentucky, 2013), 167–72; Allan M. Brandt, *Cigarette Century: The Rise, Fall, and Deadly Persistence of the Product That Defined America* (New York: Basic Books, 2007), 434–40. On the ways the TTPP changed American tobacco production, see Peter Benson, "Good, Clean Tobacco," chap. 4 in *Tobacco Capitalism: Growers, Migrant Workers, and the Changing Face of a Global Industry* (Princeton: Princeton University Press, 2012).

2. For a detailed examination of the subsistence versus surplus nature of early Appalachian agriculture, see Wilma Dunaway, "Makin' Do or Chasing Profits? The Agrarian Capitalism of Southern Appalachia," chap. 5 in *The First American Frontier: Transition to Capitalism in Southern Appalachia, 1700–1860* (Chapel Hill: University of North Carolina Press, 1996).

3. John Alexander Williams, *Appalachia: A History* (Chapel Hill: University of North Carolina Press, 2002), 91–92; David Hsiung, *Two Worlds in the Tennessee Mountains: Exploring the Origins of Appalachian Stereotypes* (Lexington: University Press of Kentucky, 1997), 76–79; John C. Inscoe, *Mountain Masters, Slavery, and the Sectional Crisis in Western*

North Carolina (Knoxville: University of Tennessee Press, 1989), 14–17; J. D. Cameron, *A Sketch of the Tobacco Interests of North Carolina* (Oxford, N.C.: W. A. Davis, 1881), 30.

4. Durwood Dunn, *Cades Cove: The Life and Death of a Southern Appalachian Community, 1818–1937* (Knoxville: University of Tennessee Press, 1988), 32–34, 76.

5. Williams, *Appalachia*, 125; Dunaway, "Makin' Do or Chasing Profits?," 141–43.

6. The classic nineteenth-century southern text on agrarian balance is John Taylor's *Arator; Being a Series of Agricultural Essays, Practical and Political* (Georgetown, D.C.: J. M. and J. B. Carter, 1813). On agricultural improvers' appeals for farm stability and diversity, good starting points include Benjamin Cohen, *Notes from the Ground: Science, Soil and Society in The American Countryside* (New Haven: Yale University Press, 2011); Emily Pawley, "'The Balance-Sheet of Nature': Calculating the New York Farm, 1820–1860" (PhD diss., University of Pennsylvania, 2009); Steven Stoll, *Larding the Lean Earth: Soil and Society in Nineteenth-Century America* (New York: Hill and Wang, 2002); and Edmund Ruffin, *Nature's Management: Writings on Landscape and Reform, 1822–1859*, ed. Jack Temple Kirby (Athens: University of Georgia Press, 2000).

7. Hsiung, *Two Worlds in the Tennessee Mountains*, esp. chaps. 3 and 5.

8. Lewis Cecil Gray, *History of Agriculture in the Southern United States to 1860* (Gloucester, Mass.: Peter Smith, 1958), 2:758; Loyal Durand Jr. and Elsie Taylor Bird, "The Burley Tobacco Region of the Mountain South," *Economic Geography* 26, no. 4 (October 1950): 281; W. H. Scherffius, H. Woosley, and C. A. Mahan, *The Cultivation of Tobacco in Kentucky and Tennessee*, USDA Farmers' Bulletin 343 (Washington D.C.: GPO, 1909), 5; J. B. Killebrew, "Report on the Culture and Curing of Tobacco in the United States," in *Report upon the Statistics of Agriculture, Compiled from Returns Received at the Tenth Census* (Washington D.C.: GPO, 1883), 167.

9. Ferrell, *Burley*, 14–19.

10. Martin Crawford, *Ashe County's Civil War: Community and Society in the Appalachian South* (Charlottesville: University Press of Virginia, 2001), 31.

11. Dunaway, *First American Frontier*, 132, 226.

12. Ibid., 245–46; Inscoe, *Mountain Masters, Slavery, and the Sectional Crisis*, 71.

13. Drew Swanson, *A Golden Weed: Tobacco and Environment in the Piedmont South* (New Haven: Yale University Press, 2014), 46–49; Barbara Hahn, *Making Tobacco Bright: Creating an American Commodity, 1617–1937* (Baltimore: Johns Hopkins University Press, 2011), 13–14.

14. Tom Lee, "Southern Appalachia's Nineteenth-Century Bright Tobacco Boom: Industrialization, Urbanization, and the Culture of Tobacco," *Agricultural History* 88, no. 2 (Spring 2014): 176; *Twelfth Census of the United States, Taken in the Year 1900: Agriculture*, pt. 2, *Crops and Irrigation* (Washington D.C.: United States Census Office, 1902), 500; Killebrew, "Report on the Culture and Curing of Tobacco," 199–200.

15. Lee, "Southern Appalachia's Nineteenth-Century Bright Tobacco Boom," 180, 182–83.

16. Cameron, *Sketch of the Tobacco Interests*, 31.

17. Samuel C. Shelton, "The Culture and Management of Tobacco," *Southern Planter* 21, no. 4 (April 1861): 209–18.

18. Samuel C. Shelton, "The Culture of Tobacco in Western North Carolina," *Southern Planter and Farmer* 3, no. 12 (December 1869): 740–41.

19. Steven E. Nash, *Reconstruction's Ragged Edge: The Politics of Postwar Life in the Southern Mountains* (Chapel Hill: University of North Carolina Press, 2016), 168–69, 172–73; Lee, "Southern Appalachia's Nineteenth-Century Bright Tobacco Boom," 179–81.

20. J. B. Killebrew and Herbert Myrick, *Tobacco Leaf: Its Culture and Cure, Marketing and Manufacture* (New York: Orange Judd, 1920), 29.

21. Swanson, *Golden Weed*, 52–55.

22. Cameron, *Sketch of the Tobacco Interests*, 33. Cameron boasted of Appalachian growers sending samples of their leaf to Constantinople to test Middle Eastern markets.

23. Killebrew, "Report on the Culture and Curing of Tobacco," 122, 125; Samuel C. Shelton, "The Culture of Tobacco in Western North Carolina," *Southern Planter and Farmer* 3, no. 12 (December 1869): 740.

24. J. B. Killebrew, "The Industrial Progress of the South," *Frank Leslie's Popular Monthly* 10, no. 6 (December 1880): 647.

25. J. B. Killebrew, "Tobacco-Discussion," *American Economic Association: Publications* 5, no. 1 (February 1904): 137; Killebrew, "Report on the Culture," 172.

26. Killebrew, "The Industrial Progress," 647.

27. Lee, "Southern Appalachia's Nineteenth-Century Bright Tobacco Boom," 185–190; Cameron, *Sketch of the Tobacco Interests*, 32, 114, 118.

28. Nash, "The Beginning of a 'New' Mountain South: Agriculture, Railroads, and Social Change, 1872–1880," chap. 6 in *Reconstruction's Ragged Edge*; Cameron, *Sketch of the Tobacco Interests*, 30; Killebrew, "Report on the Culture and Curing of Tobacco," 119.

29. Cameron, *Sketch of the Tobacco Interests*, 34–38.

30. Killebrew, "Report on the Culture and Curing of Tobacco," 124–25.

31. Lee, "Southern Appalachia's Nineteenth-Century Bright Tobacco Boom," 190–92; Katie Algeo, "The Rise of Tobacco as a Southern Appalachian Staple: Madison County, North Carolina," *Southeastern Geographer* 37, no. 1 (May 1997): 48; Cameron, *Sketch of the Tobacco Interests*, 31.

32. Killebrew, "Report on the Culture and Curing of Tobacco," 207.

33. Lee, "Southern Appalachia's Nineteenth-Century Bright Tobacco Boom," 197; Algeo, "Rise of Tobacco as a Southern Appalachian Staple," 50.

34. T. L. Copley, Clarence S. Britt, and W. B. Posey, *Conservation Practices for Tobacco Lands of the Flue-Cured and Maryland Belts*, USDA Misc. Pub. 656 (Washington D.C.: GPO, 1948), 2.

35. Soil Conservation Service, *Reconnaissance Erosion Survey of the State of North Carolina* (Washington D.C.: Soil Conservation Service, 1934); Soil Conservation Service, *Reconnaissance Erosion Survey for the State of Tennessee* (Washington D.C.: Soil Conservation Service, 1934).

36. Lee, "Southern Appalachia's Nineteenth-Century Bright Tobacco Boom," 187–88, 191–93.

37. Swanson, *Golden Weed*, 218–22; Ferrell, *Burley*, 17; Eldred E. Prince Jr. and Robert Simpson, "Pearl of the Pee Dee: 1885–1918," chap. 3 in *Long Green: The Rise and Fall of Tobacco in South Carolina* (Athens: University of Georgia Press, 2000).

38. Brandt, *Cigarette Century*, 26–43; Robert F. Durden, *Bold Entrepreneur: A Life of James B. Duke* (Durham, N.C.: Carolina Academic Press, 2003); Algeo, "Rise of Tobacco as a Southern Appalachian Staple," 51; Patrick G. Porter, "Origins of the American Tobacco Company," *Business History Review* 43, no. 1 (Spring 1969): 63–70.

39. Two of the most comprehensive retellings of this discovery narrative are Carl N. Thompson, *Historical Collections of Brown County, Ohio* (Piqua, Ohio: Hammer Graphics, 1971), 373–76; and J. B. Killebrew, *Tobacco: How to Cultivate, Cure and Prepare for Market*

(Chicago: Fertilizer Manufacturers' Association, 1900), 15–16. For a more recent treatment, see Hahn, *Making Tobacco Bright*, 78–79, 138–39.

40. Ferrell, *Burley*, 18; Lowell H. Harrison and James C. Klotter, *A New History of Kentucky* (Lexington: University Press of Kentucky, 1997), 294; W. F. Axton, *Tobacco and Kentucky* (Lexington: University Press of Kentucky, 1975), 68–69; Gray, *History of Agriculture in the Southern United States*, 2:770; Percy Wells Tidwell and John I. Falconer, *History of Agriculture in the Northern United States, 1620–1860* (Washington D.C.: Carnegie Institution, 1925); 182–83; W. H. Scherffius, H. Woosley, and C. A. Mahan, *The Cultivation of Tobacco in Kentucky and Tennessee*, USDA Farmers' Bulletin 343 (Washington D.C.: GPO, 1909), 6; Killebrew, "Report on the Culture and Curing of Tobacco," 127; *The History of Brown County, Ohio* (Chicago: W. H. Beers, 1883), 309.

41. Ferrell, *Burley*, 16; Harrison and Klotter, *New History of Kentucky*, 295; J. B. Killebrew, "Tobacco-Discussion," 138–39; Killebrew, "Report on the Culture and Curing of Tobacco," 42; J. F. B., "White Burley Tobacco: Its Cultivation," *Ohio Farmer* 90, no. 13 (September 24, 1896): 218.

42. Lee, "Southern Appalachia's Nineteenth-Century Bright Tobacco Boom," 199; Tom Lee, *The Tennessee-Virginia Tri-Cities: Urbanization in Appalachia, 1900–1950* (Knoxville: University of Tennessee Press, 2005), 164–65; Durand and Bird, "Burley Tobacco Region of the Mountain South," 284; Killebrew, *Tobacco*, 16; Killebrew, "Report on the Culture and Curing of Tobacco," 44–73.

43. "Agricultural Items," *Prairie Farmer* 44, no. 50 (December 13, 1873): 394.

44. Killebrew, "Report on the Culture and Curing of Tobacco," 6–7.

45. *Twelfth Census of the United States . . . Agriculture*, pt. 2, *Crops and Irrigation*, 505.

46. A. F. Ellis quoted in Thompson, *Historical Collections of Brown County*, 376; Robert Ragland, "Varieties of Tobacco. Soil," *American Farmer* 2, no. 5 (March 1, 1883): 69–70; "White Burley Tobacco," *Coleman's Rural World* 34, no. 16 (April 21, 1881): 124.

47. Ellis quoted in Thompson, *Historical Collections of Brown County*, 376.

48. Killebrew, *Tobacco*, 29; Killebrew, "Report on the Culture and Curing of Tobacco," 74.

49. Hahn, *Making Tobacco Bright*, 94–95, 150; Brandt, *Cigarette Century*, 23–24; "Tobacco in Virginia," *Southern Planter and Farmer* 42, no. 1 (January 1881): 69–71.

50. Killebrew, *Tobacco*, 23.

51. Hahn, *Making Tobacco Bright*, 150; Brandt, *Cigarette Century*, 26, 54, 70; Williams, *Appalachia*, 317.

52. Swanson, *Golden Weed*, chaps. 2 and 7; Algeo, "Rise of Tobacco as a Southern Appalachian Staple," 52–53.

53. Ellen Churchill Semple, "The Anglo-Saxons of the Kentucky Mountains: A Study in Anthropogeography," *Bulletin of the American Geographical Society of New York* 42, no. 1 (1910): 571–72; "Life in the Southern Appalachians," *Forest and Stream* 75, no. 16 (October 15, 1910): 610.

54. "Life in the Southern Appalachians," 611.

55. Jonathan Gilmer Speed, "The Kentucky Mountaineers," *Outlook* 48, no. 15 (October 7, 1893): 630.

56. O. E. Baker and A. B. Genung, *A Graphic Summary of Farm Crops (Based Largely on the Census of 1930 and 1935)*, USDA Misc. Pub. 267 (Washington D.C.: GPO, 1938), 9–13; Bureau of Agricultural Economics, *Economic and Social Problems and Conditions of the*

Southern Appalachians, Miscellaneous Publication 205 (Washington D.C.: GPO, 1935), 44, 48, 74–75.

57. I. S. Cook and C. H. Scherffium, *White Burley Tobacco*, West Virginia University Agricultural Experiment Station, Bulletin 152 (Morgantown: West Virginia University, 1916), 3–4.

58. Paul K. Conkin, *A Revolution Down on the Farm: The Transformation of American Agriculture since 1929* (Lexington: University Press of Kentucky, 2008), 32.

59. Lee, *Tennessee-Virginia Tri-Cities*, 179; Bureau of Agricultural Economics, *Economic and Social Problems and Conditions*, 48, 61–62, 65, 75.

60. Lee, *Tennessee-Virginia Tri-Cities*, 166.

61. Roger Biles, *The South and the New Deal* (Lexington: University Press of Kentucky, 1994), 17–18.

62. Evan P. Bennett, *When Tobacco Was King: Families, Farm Labor, and Federal Policy in the Piedmont* (Gainesville: University Press of Florida, 2014), 71–72; Biles, *South and the New Deal*, 39; Pete Daniel, *Breaking the Land: The Transformation of Cotton, Tobacco, and Rice Cultures since 1880* (Urbana: University of Illinois Press, 1985), 110–11.

63. Bennett, *When Tobacco Was King*, 71–72; Ferrell, *Burley*, 22–24; Conkin, *Revolution Down on the Farm*, 67–68.

64. Benson, *Tobacco Capitalism*, 81; Lee, *Tennessee-Virginia Tri-Cities*, 178; Daniel, *Breaking the Land*, 116–20.

65. Bennett, *When Tobacco Was King*, 74; Conkin, *Revolution Down on the Farm*, 88; Harrison and Klotter, *New History of Kentucky*, 295; Biles, *South and the New Deal*, 41.

66. Durand and Bird, "Burley Tobacco Region of the Mountain South," 287.

67. Quoted in Randall Norris and Jean-Philippe Cypres, *Women of Coal* (Lexington: University Press of Kentucky, 1996), 38.

68. Conkin, *Revolution Down on the Farm*, 68; Jerry Thomas, *An Appalachian New Deal: West Virginia in the Great Depression* (Lexington: University Press of Kentucky, 1998), 166.

69. Lee, *Tennessee-Virginia Tri-Cities*, 179–80.

70. Bennett, *When Tobacco Was King*, 71; Harrison and Klotter, *New History of Kentucky*, 295; Bureau of Agricultural Economics, *A Graphic Summary of Farm Crops (Based Largely on the Census of 1940)*, USDA Misc. Pub. 512 (Washington D.C.: GPO, 1943), 11.

71. Neil M. Maher, *Nature's New Deal: The Civilian Conservation Corps and the Roots of the American Environmental Movement* (New York: Oxford University Press, 2008), 63–64; Williams, *Appalachia*, 317.

72. Hahn, in *Making Tobacco Bright*, demonstrates ways tobacco culture has been "naturalized."

73. Mark A. Swanson, "No Substitute for Tobacco: The Search for Farm Diversification in Appalachian Kentucky" (PhD diss., University of Florida, 2001), 91–92; Algeo, "Rise of Tobacco as a Southern Appalachian Staple," 53–54.

74. Sara Gregg, "Uncovering the Subsistence Economy in the Twentieth Century South: Blue Ridge Mountain Farms," *Agricultural History* 78, no. 4 (2004): 421–22.

75. Jenrose Fitzgerald, Lisa Markowitz, and Dwight B. Billings, "Not Your Grandmother's Agrarianism: The Community Farm Alliance's Agrifood Activism," in *Transforming Places: Lessons from Appalachia*, ed. Stephen L. Fisher and Barbara Ellen Smith (Urbana: University of Illinois Press, 2012), 210–25, quote on 210.

Chapter 8. Power: Building an Atomic Appalachia in East Tennessee

1. George O. Robinson Jr., *The Oak Ridge Story: The Saga of a People Who Share in History* (Kingsport, Tenn.: Southern Publishers, 1950), 17–19. The Hendrix story is repeated in Oak Ridge National Laboratory's institutional history, Oak Ridge National Laboratory (hereafter ORNL), "Swords to Plowshares: A Short History of Oak Ridge National Laboratory (1943–1993)," U.S. Department of Energy, Office of Science and Technical Information, https://www.osti.gov/opennet/manhattan-project-history/publications /SwordstoPlowshares-AShortHistoryofORNL.pdf, as well as Denise Kiernan, *The Girls of Atomic City: The Untold Story of the Women Who Helped Win World War II* (New York: Touchstone, 2013), 27–28.

2. Some interesting introductory sources on the Manhattan Project include Richard Rhodes, *The Making of the Atomic Bomb* (New York: Simon & Schuster, 1986); Cynthia C. Kelly, ed., *The Manhattan Project: The Birth of the Atomic Bomb in the Words of Its Creators, Eyewitnesses, and Historians* (New York: Black Dog & Leventhal, 2009); Richard G. Hewlett and Oscar E. Anderson, *The New World, 1939–1946: A History of the United States Atomic Energy Commission* (Washington, D.C.: U.S. Atomic Energy Commission, 1972); Leslie R. Groves, *Now It Can Be Told: The Story of the Manhattan Project* (New York: Harper & Brothers, 1962); Ray Monk, *Robert Oppenheimer: A Life inside the Center* (New York: Anchor, 2014); and Kai Bird and Martin J. Sherwin, *American Prometheus: The Triumph and Tragedy of J. Robert Oppenheimer* (New York: Vintage Books, 2005).

3. John Alexander Williams, *Appalachia: A History* (Chapel Hill: University of North Carolina Press, 2002), 366.

4. Robinson, *Oak Ridge Story*, 91.

5. Peter Bacon Hales, *Atomic Spaces: Living on the Manhattan Project* (Urbana: University of Illinois Press, 1997), 12; Laurence C. Walker, *The Southern Forest: A Chronicle* (Austin: University of Texas Press, 1991), 22, 176; Leland Johnson and Daniel Schaffer, *Oak Ridge National Laboratory: The First Fifty Years* (Knoxville: University of Tennessee Press, 1994), 1; Robinson, *Oak Ridge Story*, 22, 26.

6. Aaron D. Purcell, *Arthur Morgan: A Progressive Vision for American Reform* (Knoxville: University of Tennessee Press, 2014); Sarah T. Phillips, *This Land, This Nation: Conservation, Rural America, and the New Deal* (New York: Cambridge University Press, 2007), 83–107; Kenneth M. Murchison, *The Snail Darter Case: TVA versus the Endangered Species Act* (Lawrence: University Press of Kansas, 2007), 8–12; Williams, *Appalachia*, 291–95; Margaret Lynn Brown, "A Lake in the National Defense," chap. 5 in *The Wild East: A Biography of the Great Smoky Mountains* (Gainesville: University Press of Florida, 2000); Albert Cowdrey, *This Land, This South: An Environmental History*, rev. ed. (Lexington: University Press of Kentucky, 1996), 152–56.

7. Phillips, *This Land, This Nation*, 83.

8. Johnson and Schaffer, *Oak Ridge National Laboratory*, 1–2; Robinson, *Oak Ridge Story*, 22, quote on 26.

9. Robinson, *Oak Ridge Story*, 25.

10. Hales, *Atomic Spaces*, 51; Johnson and Schaffer, *Oak Ridge National Laboratory*, 16.

11. The classic example is Mandel Sherman and Thomas Henry, *Hollow Folk* (New York: Thomas Y. Crowell, 1933), a sociological study of Shenandoah communities used to justify the dispossession of residents within the boundaries of the new Shenandoah National Park.

12. Quoted in Charles L. Perdue Jr. and Nancy J. Martin-Perdue, "Appalachian Fables and Facts: A Case Study of the Shenandoah National Park Removals," *Appalachian Journal* 7, nos. 1–2 (Autumn/Winter 1979/1980): 89.

13. Katrina Powell, *"Answer at Once": Letters of Mountain Families in Shenandoah National Park, 1934–1938* (Charlottesville: University of Virginia Press, 2009), 4–9; Anne Mitchell Whisnant, *Super-Scenic Motorway: A Blue Ridge Parkway History* (Chapel Hill: University of North Carolina Press, 2006), chaps. 3–5; Terry Young, "False, Cheap and Degraded: When History, Economy and Environment Collided at Cades Cove, Great Smoky Mountains National Park," *Journal of Historical Geography* 32, no. 1 (2006): 169–189; Margaret Brown, "Scenery in the Eminent Domain," chap. 3 in *The Wild East: A Biography of the Great Smoky Mountains* (Gainesville: University Press of Florida, 2000); Dan Pierce, "The Barbarism of the Huns: Family and Community Removal in the Establishment of the Great Smoky Mountains National Park," *Tennessee Historical Quarterly* 57, no. 1 (1998): 62–79; Phil Noblitt, "The Blue Ridge Parkway and Myths of the Pioneer," *Appalachian Journal* 21, no. 4 (Summer 1994): 394–408; Durwood Dunn, "Death by Eminent Domain," chap. 10 in *Cades Cove: The Life and Death of a Southern Appalachian Community, 1818–1937* (Knoxville: University of Tennessee Press, 1988).

14. Sara Gregg, *Managing the Mountains: Land Use Planning, the New Deal, and the Creation of a Federal Landscape in Appalachia* (New Haven: Yale University Press, 2010); "The Depression and the New Deal," chap. 3 in *U.S. Forest Service History*, Forest History Society, https://web.archive.org/web/20160216182927/http://www.foresthistory.org/ASPNET/Publications/region/8/history/chap3.aspx.

15. Phillips, *This Land, This Nation*, 120–28; Rexford Tugwell, "The Resettlement Administration," *Agricultural History* 33 (1959): 160.

16. "The Depression and the New Deal."

17. Michael J. McDonald and John Muldowny, *TVA and the Dispossessed: The Resettlement of Population in the Norris Dam Area* (Knoxville: University of Tennessee Press, 1981); Neil M. Maher, *Nature's New Deal: The Civilian Conservation Corps and the Roots of the American Environmental Movement* (New York: Oxford University Press, 2008), 191–95; Phillips, *This Land, This Nation*, 83–107.

18. Robinson, *Oak Ridge Story*, 26, 28; Kiernan, *Girls of Atomic City*, 25.

19. Eula and Beulah Cooper, interviewed by Jennifer Thonhoff, April 15, 2005, Center for Oak Ridge Oral History, http://cdm16107.contentdm.oclc.org/cdm/singleitem/collection/p15388coll1/id/91/rec/71; Bonita Irwin, interviewed by Jennifer Thonhoff, September 23, 2005, Center for Oak Ridge Oral History, http://cdm16107.contentdm.oclc.org/cdm/singleitem/collection/p15388coll1/id/101/rec/153. The Center for Oak Ridge Oral History (hereafter COROH) has collected and made available online accounts from dozens of former plant workers and town residents.

20. Warren Bradshaw, interviewed by Bart Callan, May 19, 2005, COROH, http://cdm16107.contentdm.oclc.org/cdm/singleitem/collection/p15388coll1/id/87/rec/39.

21. Hales, *Atomic Spaces*, 47–57.

22. Irwin interview.

23. Eula and Beulah Cooper interview.

24. Ralph Aurin, interviewed by Don Hunnicutt, December 11, 2012, COROH, http://cdm16107.contentdm.oclc.org/cdm/singleitem/collection/p15388coll1/id/386/rec/6; Mary Elizabeth Alexander, interviewed by Don Hunnicutt, December 19, 2012, COROH, http://cdm16107.contentdm.oclc.org/cdm/singleitem/collection/p15388coll1/id/376/rec/3.

25. Horace Stanley, interviewed by Bart Callan, April 12, 2005, COROH, http://cdm16107
.contentdm.oclc.org/cdm/singleitem/collection/p15388coll1/id/126/rec/7.

26. Dorothy Gilpatrick, unknown interviewer, July 2001, COROH, http://cdm16107
.contentdm.oclc.org/cdm/singleitem/collection/p15388coll1/id/330/rec/113.

27. Kelly, *Manhattan Project*, 201–2.

28. Johnson and Schaffer, *Oak Ridge National Laboratory*, 2–3; Groves, *Now It Can Be Told*, 94–95, 111; John Purcell, *The Best-Kept Secret: The Story of the Atomic Bomb* (New York: Vanguard, 1963), 127–28.

29. Adam Rome, "Levitt's Progress: The Rise of the Suburban-Industrial Complex," chap. 1 in *The Bulldozer in the Countryside: Suburban Sprawl and the Rise of American Environmentalism* (Cambridge, UK: Cambridge University Press, 2001); Robinson, *Oak Ridge Story*, 61–62; ORNL, "Swords to Plowshares"; Alexander interview. On the evolution of suburban forms in general, see Kenneth T. Jackson, *Crabgrass Frontier: The Suburbanization of the United States* (New York: Oxford University Press, 1985).

30. Stuart Patterson, "A Brave New Deal World: The Cumberland Homesteads," in *Rural Life and Culture in the Upper Cumberland*, ed. Michael E. Birdwell and W. Calvin Dickinson (Lexington: University Press of Kentucky, 2004): 197–98.

31. Johnson and Schaffer, *Oak Ridge National Laboratory*, 3.

32. Purcell, *Best-Kept Secret*, 123.

33. Michael C. C. Adams, *The Best War Ever: America and World War II*, 2nd ed. (Baltimore: Johns Hopkins University Press, 2015), 90.

34. Hales, *Atomic Spaces*, 92–93.

35. Ibid., 79.

36. Aurin interview.

37. Ruby Shanks, interviewed by Bart Callan, May 18, 2005, COROH, http://cdm16107
.contentdm.oclc.org/cdm/singleitem/collection/p15388coll1/id/124/rec/399.

38. Jane Richardson, interviewed by Don Hunnicutt, January 21, 2014, COROH, http://
cdm16107.contentdm.oclc.org/cdm/singleitem/collection/p15388coll1/id/496/rec/362.

39. Joanne Gailar, interviewed by Bart Callan, March 7, 2005, COROH, http://cdm16107
.contentdm.oclc.org/cdm/singleitem/collection/p15388coll1/id/95/rec/15.

40. Poem by Oak Ridge resident Colleen Black, in Kelly, *Manhattan Project*, 202.

41. Theodore Rockwell quoted in ibid., 208.

42. Evelyn Ellingson, interviewed by Barbara P. Campbell, August 21, 2002, COROH, http://cdm16107.contentdm.oclc.org/cdm/singleitem/collection/p15388coll1/id/324/rec/86.

43. Margene Lyon, interviewed by Connie Callan, May 17, 2005, COROH, http://
cdm16107.contentdm.oclc.org/cdm/singleitem/collection/p15388coll1/id/111/rec/212.

44. Aurin interview. COROH contains numerous interviews that include recollections of the boardwalks.

45. James Hackworth, interviewed by Bart Callan, April 13, 2005, COROH, http://cdm
16107.contentdm.oclc.org/cdm/singleitem/collection/p15388coll1/id/97/rec/121.

46. Groves, *Now It Can Be Told*, 106.

47. Quoted in Kelly, *Manhattan Project*, 204.

48. Theodore Rockwell, interview by Cynthia C. Kelly, Atomic Heritage Foundation, "Voices of the Manhattan Project," Oak Ridge, Tenn., April 27, 2002, http://
manhattanprojectvoices.org/oral-histories/theodore-rockwells-interview-2002.

49. ORNL, "Swords to Plowshares."

50. Kiernan, *Girls of Atomic City*, 151–55; Hales, *Atomic Spaces*, 92–93, 123–24, 128, quote on 119.

51. Purcell, *Best-Kept Secret*, 128–29; Robinson, "They Couldn't Say a Word: How the Secrecy Was Maintained on the Most Secret of Projects," chap. 5 in *Oak Ridge Story*; quote in Gailar interview.

52. Robert Allen, interviewed by Chris Albrecht, November 3, 2005, COROH, http://cdm16107.contentdm.oclc.org/cdm/singleitem/collection/p15388coll1/id/413/rec/4; Kelly, *Manhattan Project*, 212–13.

53. Hales, *Atomic Spaces*, 80–83.

54. Louise Alspaugh, interviewed by Jim Kolb, Sept. 17, 2002, COROH, http://cdm16107 .contentdm.oclc.org/cdm/singleitem/collection/p15388coll1/id/21/rec/5. On the gate opening, see also Mary Elizabeth Alexander, Interviewed by Don Hunnicutt, December 19, 2012, COROH, http://cdm16107.contentdm.oclc.org/cdm/singleitem/collection/p15388coll1 /id/376/rec/3.

55. Lindsey A. Freeman, *Longing for the Bomb: Oak Ridge and Atomic Nostalgia* (Chapel Hill: University of North Carolina Press, 2015). On the tension between security and safety in everyday Oak Ridge life, see Charles W. Johnson and Charles O. Jackson, *City behind a Fence: Oak Ridge, Tennessee, 1942–1946* (Knoxville: University of Tennessee Press, 1981).

56. Freeman, *Longing for the Bomb*, 66–69, 76–77, 89–90. On the varied dangers of work in the complex as well as workers' efforts to understand and protect themselves against these hazards (especially in the postwar period), see Russell Olwell, *At Work in the Atomic City: A Labor and Social History of Oak Ridge, Tennessee* (Knoxville: University of Tennessee Press, 2004).

57. Earline Banic, interviewed by Don Hunnicutt, October 31, 2012, COROH, http://cdm16107.contentdm.oclc.org/cdm/singleitem/collection/p15388coll1/id/337/rec/13.

58. Kate Brown, *Dispatches from Dystopia: Histories of Places Not Yet Forgotten* (Chicago: University of Chicago Press, 2015), 75.

59. Mark Fiege, *The Republic of Nature: An Environmental History of the United States* (Seattle: University of Washington Press, 2012), 292.

60. These representative statistics can be found in Groves, *Now It Can Be Told*, 97, 99, 103–4, 113–15; Robinson, *Oak Ridge Story*, 79–80, 83.

61. ORNL, "Swords to Plowshares."

62. Groves, *Now It Can Be Told*, 112, 114; Johnson and Schaffer, *Oak Ridge National Laboratory*, 17–18.

63. Groves, *Now It Can Be Told*, 107–9.

64. Ibid., 114–15.

65. Ibid., 107–9.

66. Lloyd Blackwood, with wife, Barbara Blackwood, interviewed by Bill Wilcox, April 6, 2011, COROH, http://cdm16107.contentdm.oclc.org/cdm/singleitem/collection /p15388coll1/id/62/rec/35.

67. Robinson, *Oak Ridge Story*, 66–67.

68. Gailer interview.

69. Stanley interview.

70. Johnson and Schaffer, "High-Flux Years," chap. 2 in *Oak Ridge National Laboratory*; Robinson, *Oak Ridge Story*, 114.

71. Randal L. Hall, *Mountains on the Market: Industry, the Environment, and the South* (Lexington: University Press of Kentucky, 2012), 121–22.

72. Suzanne Marshall, Rufus Kinney, and Antoinette Hudson, "The Incineration of Chemical Weapons in Anniston, Alabama: The March for Environmental Justice," in *Confronting Ecological Crises in Appalachia and the South: University and Community Partnerships*, ed. Stephanie McSpirit, Lynne Faltraco, and Conner Bailey (Lexington: University Press of Kentucky, 2012), 147–48. For the ensuing environmental and health problems in Anniston, see Ellen Griffith Spears, *Baptized in PCBs: Race, Pollution, and Justice in an All-American Town* (Chapel Hill: University of North Carolina Press, 2014).

73. Ronald D. Eller, *Uneven Ground: Appalachia since 1945* (Lexington: University Press of Kentucky, 2008), 84–87, 104–5, 110–14.

74. Ibid., 177–178, 187.

75. For insightful studies of the nature of atomic sites, see Hale, *Atomic Spaces*; Freeman, *Longing for the Bomb*; Kate Brown, *Plutopia: Nuclear Families, Atomic Cities, and the Great Soviet and American Plutonium Disasters* (New York: Oxford University Press, 2013); and Richard White, "The Power of the River," chap. 3 in *The Organic Machine: The Remaking of the Columbia River* (New York: Hill & Wang, 1995).

76. An exception is Williams, *Appalachia*, 313, 317, 366.

77. Freeman, *Longing for the Bomb*, 15–16.

Chapter 9. Coal: Sludge Ponds and Vanishing Mountains

1. Harry M. Caudill, *Night Comes to the Cumberlands: A Biography of a Depressed Area* (Boston: Little, Brown, 1963), x.

2. Stephanie McSpirit, Sharon Hardesty, Patrick Carter-North, Mark Grayson, and Nina McCoy, "The Martin County Project: Students, Faculty, and Citizens Research the Effects of a Technological Disaster," in *Confronting Ecological Crises in Appalachia and the South: University and Community Partnerships*, ed. Stephanie McSpirit, Lynne Faltraco, and Conner Bailey (Lexington: University Press of Kentucky, 2012), 761; "POLREPS [pollution report]," October 16 and October 17, 2000, Environmental Protection Agency, Region 4 (hereafter EPA). This and other EPA documents concerning the Martin County spill and agency responses can be found on the EPA's website at http://www.epa.gov/region4/foiapgs/readingroom/martincoal/. Figures for the Gulf spill come from National Response Team, *On Scene Coordinator Report:* Deepwater Horizon *Oil Spill* (Washington, D.C.: U.S. Coast Guard, 2011). The *Deepwater Horizon* spill totaled 4.9 million barrels, at forty-two gallons per barrel, with an estimate error rate of plus or minus 10 percent (p. 33).

3. Shaunna L. Scott, Stephanie McSpirit, Sharon Hardesty, and Robert Welch, "Post Disaster Interviews with Martin County Citizens: 'Gray Clouds' of Blame and Distrust," *Journal of Appalachian Studies* 11, 1–2 (Spring/Fall 2005): 7–8; Peter T. Kilborn, "A Torrent of Sludge Muddies a Town's Future," *New York Times*, December 25, 2000.

4. Kai T. Erikson, *Everything in Its Path: Destruction of Community in the Buffalo Creek Flood* (New York: Simon & Schuster, 1978); Gerald M. Stern, *The Buffalo Creek Disaster: The Story of the Survivors' Unprecedented Lawsuit* (New York: Random House, 1976).

5. Cassie Robinson Pfleger, Randal Pfleger, Ryan Wishart, and Dave Cooper, "Mountain Justice," in *Transforming Places: Lessons from Appalachia*, ed. Stephen L. Fisher and Barbara Ellen Smith (Urbana: University of Illinois Press, 2012), 226–38; Anita Puckett, Elizabeth Fine, Mary Hufford, Ann Kingsolver, and Betsy Taylor, "Who Knows? Who Tells? Creating a Knowledge Commons," in Fisher and Smith, *Transforming Places*, 239–

51; Talmage A. Stanley, *The Poco Field: An American Story of Place* (Urbana: University of Illinois Press, 2012), 113–14; *The Last Mountain* (Docurama Films, 2011).

6. Ronald Eller, *Miners, Millhands, and Mountaineers: Industrialization of the Appalachian South, 1880–1930* (Knoxville: University of Tennessee Press, 1982), 54–58; Chad Montrie, *To Save the Land and People: A History of Opposition to Surface Coal Mining in Appalachia* (Chapel Hill: University of North Carolina Press, 2003), 66–67.

7. Montrie, *To Save the Land and People*, 17–20; Robert F. Munn, "The Development of Strip Mining in Southern Appalachia," *Appalachian Journal* 3, no. 1 (Autumn 1975): 87–88.

8. Ted Steinberg, *Down to Earth: Nature's Role in American History*, 2nd ed. (New York: Oxford University Press, 2009), 219–20; Adam Rome, *The Bulldozer in the Countryside: Suburban Sprawl and the Rise of American Environmentalism* (Cambridge: Cambridge University Press, 2001), esp. chap. 2; David Nye, *Consuming Power: A Social History of American Energies* (1998; repr., Cambridge: MIT Press, 1999), 204–5.

9. Caudill, *Night Comes to the Cumberlands*, 258–62; Ronald D. Eller, *Uneven Ground: Appalachia since 1945* (Lexington: University Press of Kentucky, 2008), 14, 16, 18–20.

10. Montrie, *To Save the Land and People*, 17–24, quote on 24; Munn, "Development of Strip Mining," 88–90; Eller, *Uneven Ground*, 36–38.

11. Eller, *Uneven Ground*, 20; Jack Weller, *Yesterday's People: Life in Contemporary Appalachia* (Lexington: University of Kentucky Press, 1966), 19–20.

12. Eller, *Uneven Ground*, 38–39.

13. Montrie, *To Save the Land and People*, 67.

14. Peter Crow, *Do, Die, or Get Along: A Tale of Two Appalachian Towns* (Athens: University of Georgia Press, 2007), 85.

15. Randall Norris and Jean-Philippe Cypres, *Women of Coal* (Lexington: University Press of Kentucky, 1996), 78.

16. Montrie, *To Save the Land and People*, 66–67.

17. Caudill, *Night Comes to the Cumberlands*, 307.

18. Ibid.; Harry M. Caudill, *My Land Is Dying* (Boston, Mass.: E. P. Dutton, 1971); Eller, *Uneven Ground*, 66–67.

19. Chad Montrie, "'We Mean to Stop Them, One Way or Another': Coal, Power, and the Fight against Strip Mining in Appalachia," in *Mountains of Injustice: Social and Environmental Justice in Appalachia*, ed. Michele Morrone and Geoffrey L. Buckley (Athens: Ohio University Press, 2011), 82–89; Chad Montrie, *To Save the Land and People*, 1, 87–88.

20. Crow, *Do, Die, or Get Along*, 85–89; Montrie, "We Mean to Stop Them," 84.

21. Silas House and Jason Howard, *Something's Rising: Appalachians Fighting Mountaintop Removal* (Lexington: University Press of Kentucky, 2009), 2–3.

22. Montrie, "We Mean to Stop Them," 83, 91.

23. House and Howard, *Something's Rising*, 2–3; Crow, *Do, Die, or Get Along*, 94–96.

24. House and Howard, *Something's Rising*, 2–3; Shirley Stewart Burns, "Mountaintop Removal in Central Appalachia," *Southern Spaces*, September 30, 2009, http://southernspaces.org/2009/mountaintop-removal-central-appalachia.

25. Quoted in Geoffrey L. Buckley and Laura Allen, "Stories about Mountaintop Removal in the Appalachian Coalfields," in Morrone and Buckley, *Mountains of Injustice*, 165; Michael Shnayerson, *Coal River: How a Few Brave Americans Took on a Powerful Company—and the Federal Government—to Save the Land They Love* (New York: Farrar, Straus and Giroux, 2008), 118–25.

26. Wendell Berry, *The Gift of Good Land: Further Essays Cultural and Agricultural* (Berkeley, Calif.: Counterpoint, 1981), 203, 204.

27. Quoted in Buckley and Allen, "Stories about Mountaintop Removal," 165.

28. Montrie, "We Mean to Stop Them," 91; Scott et al., "Post Disaster Interviews with Martin County Citizens," 16; Kilborn, "Torrent of Sludge."

29. Susan F. Hirsch and Franklin Dukes, *Mountaintop Mining in Appalachia: Understanding Stakeholders and Change in Environmental Conflict* (Athens: Ohio University Press, 2014), 12.

30. Erik Reece and James Krupa, *The Embattled Wilderness: The Natural and Human History of Robinson Forest and the Fight for Its Future* (Athens: University of Georgia Press, 2013), 20–21; House and Howard, *Something's Rising*, 2–4.

31. "POLREPS," October 17, 2000, EPA; "POLREPS," November 17, 2000, EPA.

32. "POLREPS," October 19, 2000, EPA; "POLREPS," October 25, 2000, EPA; "POLREPS," December 21, 2000, EPA.

33. "POLREPS," December 21, 2000, EPA, quote on 2.

34. *Sludge*, dir. Robert Salyer (Appalshop, 2005).

35. House and Howard, *Something's Rising*, 182, 197–98, quote on 192.

36. Ibid., 181–84, 192–94; Ted Williams, "Sludge Slinging," *Audubon* 106, no. 2 (May 2004): 22–31.

37. Quoted in House and Howard, *Something's Rising*, 63.

38. Quoted in ibid., 198.

39. "The Poverty Tours (April–May 1964)," Lyndon Baines Johnson Library, Government Film MP791, available on the library's YouTube channel, https://www.youtube.com/watch?v=oVyZ_vKuY-M. For contemporary local understandings of Johnson's Martin County visit, see Pam Fessler, "Kentucky County That Gave War on Poverty a Face Still Struggles," NPR, www.npr.org/2014/01/08/260151923/kentucky-county-that-gave-war-on-poverty-a-face-still-struggles.

40. Shnayerson, *Coal River*, 8.

41. Scott et al., "Post Disaster Interviews with Martin County Citizens," 13–25; Stephanie McSpirit, Shaunna L. Scott, Sharon Hardesty, and Robert Welch, "EPA Actions in Post Disaster Martin County, Kentucky: An Analysis of Bureaucratic Slippage and Agency Recreancy," *Journal of Appalachian Studies* 11, nos. 1–2 (Spring/Fall 2005): 30–63.

42. Scott et al., "Post Disaster Interviews with Martin County Citizens," 14.

43. Ibid., 16. For the long history of American efforts to blame calamities on Providence, see Ted Steinberg, *Acts of God: The Unnatural History of Natural Disaster in America* (New York: Oxford University Press, 2000).

44. Hirsch and Dukes, *Mountaintop Mining in Appalachia*, 22–23; Shnayerson, *Coal River*, 128–45; House and Howard, *Something's Rising*, 10; "Classes Begin at New Marsh Fork Elementary School," January 9, 2013, http://ilovemountains.org/news/3566.

45. Rebecca Scott, *Removing Mountains: Extracting Nature and Identity in the Appalachian Coalfields* (Minneapolis: Quadrant/University of Minnesota Press, 2010); Joyce Barry, *Standing Our Ground: Women, Environmental Justice, and the Fight to End Mountaintop Removal* (Athens: Ohio University Press, 2012).

46. Montrie, *To Save the Land and People*; Shirley Stewart Burns, *Bringing down the Mountains: The Impact of Mountaintop Removal on West Virginia Communities* (Morgantown: West Virginia University Press, 2007).

47. For example, see Stephanie McSpirit, Sharon Hardesty, Patrick Carter-North, Mark Grayson, and Nina McCoy, "The Martin County Project: Students, Faculty, and Citizens Research the Effects of a Technological Disaster," in *Confronting Ecological Crises in Appalachia and the South: University and Community Partnerships*, ed. Stephanie McSpirit, Lynne Faltraco, and Conner Bailey (Lexington: University Press of Kentucky, 2012), 75–91; and Robert Gipe, "Unsuitable: The Fight to Save Black Mountain, 1998–1999," in ibid., 93–107.

48. Erik Reece, *Lost Mountain: A Year in the Vanishing Wilderness: Radical Strip Mining and the Devastation of Appalachia* (New York: Riverhead Books, 2006).

49. Ann Pancake, *Strange as This Weather Has Been* (San Francisco: Shoemaker and Hoard, 2007).

50. Shnayerson, *Coal River*.

51. Cassie Robinson Pfleger, Randal Pfleger, Ryan Wishart, and Dave Cooper, "Mountain Justice," in Fisher and Smith, *Transforming Places*, 226–38.

52. *Sludge*, dir. Robert Salyer (Appalshop, 2005).

53. *The Last Mountain*, dir. Bill Haney (Massachusetts Documentary Productions/Uncommon Productions, 2011).

54. iLoveMountainsOrg, YouTube.org, www.youtube.com/user/iLoveMountainsOrg /videos.

55. Morrone and Buckley, *Mountains of Injustice*, xi; Burns, "Mountaintop Removal in Central Appalachia."

56. "Dan River Coal Ash Spill Damage Could Top $300 Million," *Raleigh News and Observer*, November 26, 2014. See also "Dan River Coal Ash Spill," collected government documents regarding the Dan River spill, North Carolina Department of Environmental Quality, http://danriverspill.ncdenr.gov.

57. On the long history and recent development of the Marcellus shale, see Tom Wilber, *Under the Surface: Fracking, Fortunes, and the Fate of the Marcellus Shale* (Ithaca, N.Y.: Cornell University Press, 2012); David A. Waples, "Back to the Future: Appalachia and the Marcellus Shale," chap. 9 in *The Natural Gas Industry in Appalachia: A History from the First Discovery to the Tapping of the Marcellus Shale*, 2nd ed. (Jefferson, N.C.: McFarland, 2012). Journalist Jeff Goodell popularized the phrase "national sacrifice zone" in *Big Coal: The Dirty Secret behind America's Energy Future* (Boston: Houghton Mifflin Harcourt, 2006).

58. Samuel P. Hays, *Beauty, Health, and Permanence: Environmental Politics in the United States, 1955–1985* (Cambridge, UK: Cambridge University Press, 1989).

59. For a range of contemporary efforts to use the Internet to form activist communities within and beyond Appalachia, see Fisher and Smith, eds., *Transforming Places*.

60. Jeff Biggers, *The United States of Appalachia: How Southern Mountaineers Brought Independence, Culture, and Enlightenment to America* (New York: Shoemaker and Hoard, 2006), 159–60; Lon Savage, *Thunder in the Mountains: The West Virginia Mine War, 1920–1921* (Pittsburgh: University of Pittsburgh Press, 1989).

61. Rachel Donaldson, "Placing and Preserving Labor History," *Public Historian* 39, no. 1 (February 2017): 80–82; Susan F. Hirsch and Franklin Dukes, *Mountaintop Mining in Appalachia: Understanding Stakeholders and Change in Environmental Conflict* (Athens: Ohio University Press, 2014), 16; Kate Sheppard, "The Battle of Blair Mountain, Round Two," *Mother Jones*, www.motherjones.com/environment/2010/11/massey-arch-coal-blair -mountain; Paul J. Nyden, "Federal Appeals Court Says Groups Can Sue over Blair Mountain," *Charleston (W.Va.) Gazette-Mail*, August 26, 2014.

Epilogue. The Adelgid and the Salamander

1. Alfred Crosby popularized the phrase in *The Columbian Exchange: Biological and Cultural Consequences of 1492* (Westport, Conn.: Greenwood, 1973).

2. According to an H-Net post by Dale Goble, a law professor at the University of Idaho, on H-Environment, November 19, 2009, the use of "charismatic megafauna" to describe large animals popular with the general public dates to the mid-1980s and appeared more frequently during the 1990s. See http://h-net.msu.edu/cgi-bin/logbrowse.pl?trx=vx& list=h-environment&month=0911&week=c&msg=uUInEpxIGOCnIKVv/Do8vA&user= &pw=.

3. Steve Nash, *Blue Ridge 20/20: An Owner's Manual* (Chapel Hill: University of North Carolina Press, 1999), 133–34; Miles Tager, *Grandfather Mountain: A Profile* (Boone, N.C.: Parkway Publishers, 1999), 29–32.

4. On various manifestations of imperial collecting during the colonial era, see Beth Fowkes Tobin, *The Duchess's Shells: Natural History Collecting in the Age of Cook's Voyages* (New Haven: Yale University Press, 2014); Londa Schiebinger, *Plants and Empire: Colonial Bioprospecting in the Atlantic World* (Cambridge: Harvard University Press, 2004); Richard Drayton, *Nature's Government: Science, Imperial Britain, and the 'Improvement' of the World* (New Haven: Yale University Press, 2000); Ray Desmong, *Kew: A History* (London: Harvill Press, 1998); Richard Grove, *Green Imperialism: Colonial Expansion, Tropical Island Edens and the Origins of Environmentalism, 1600–1860* (Cambridge, UK: Cambridge University Press, 1996).

5. See also Drew A. Swanson, "Endangered Species and Threatened Landscapes in Appalachia: Managing the Wild and the Human in the American Mountain South," *Environment and History* 18, no. 1 (February 2012): 44–45 (esp. note 30).

6. On Weller, see Charles E. Burt, "A Contribution to the Herpetology of Kentucky," *American Midland Naturalist* 14, no. 6 (November 1933): 669–79; Tager, *Grandfather Mountain*, 30–31; Ellin Beltz, "Biographies of People Honored in the Names of the Reptiles and Amphibians of North America," www.ebeltz.net/herps/biogappx.html#W.

7. Kathryn Newfont, *Blue Ridge Commons: Environmental Activism and Forest History in Western North Carolina* (Athens: University of Georgia Press, 2012), 39–42; Jack Temple Kirby, *Mockingbird Song: Ecological Landscapes of the South* (Chapel Hill: University of North Carolina Press, 2006), 150; Ralph H. Lutts, "Like Manna from God: The American Chestnut Trade in Southwestern Virginia," *Environmental History* 9, no. 3 (July 2004): 497–525; Margaret Lynn Brown, *The Wild East: A Biography of the Great Smoky Mountains* (Gainesville: University Press of Florida, 2000), 99–102; Donald Davis, *Where There Are Mountains: An Environmental History of the Southern Appalachians* (Athens: University of Georgia Press, 2000), 192–98.

8. Newfont, *Blue Ridge Commons*, 39.

9. Davis, *Where There Are Mountains*, 192.

10. Timothy Silver, *Mount Mitchell and the Black Mountains: An Environmental History of the Highest Peaks in Eastern America* (Chapel Hill: University of North Carolina Press, 2003), 236–38; N. S. Nicholas, S. M. Zedaker, and C. Eager, "A Comparison of Overstory Community Structure in Three Southern Appalachian Spruce-Fir Forests," *Bulletin of the Torrey Botanical Club* 119, no. 3 (July–September 1992): 318.

11. Silver, *Mount Mitchell*, 236–37.

12. Ibid., 238.

13. Nash, *Blue Ridge 20/20*, 27–28, 63.

14. Michael A. Jenkins, "Impact of the Balsam Woolly Adelgid (*Adelges piceae* Ratz.) on an *Abies fraseri* (Pursh) Poir. Dominated Stand Near the Summit of Mount LeConte, Tennessee," *Castanea* 68, no. 2 (June 2003): 116.

15. James A. Organ, "Studies on the Life History of the Salamander, *Plethodon welleri*," *Copeia* 1960, no. 4 (December 30, 1960): 294.

16. Silver, *Mount Mitchell*, 240–43.

17. Ibid.; Nicholas, Zedaker, and Eager, "Comparison of Overstory," 318; Nash, *Blue Ridge 20/20*, 26.

18. Katherine Lee Martin, "Ecosystem Dynamics in Central Appalachian Riparian Forests Affected by Hemlock Woolly Adelgid" (PhD diss., Ohio State University, 2012), 6; Nash, *Blue Ridge 20/20*, 30.

19. Mark Faulkenberry, Roy Hedden, and Joe Culin, "Hemlock Susceptibility to Hemlock Woolly Adelgid Attack in the Chattooga River Watershed," *Southeastern Naturalist* 8, no. 1 (March 1, 2009): 129, quote on 138. For one journalist's engaging narrative of HWA's spread, see Richard Preston, "A Death in the Forest," chap. 2 in *Panic in Level 4: Cannibals, Killer Viruses, and Other Journeys to the Edge of Science* (New York: Random House, 2009).

20. Robert M. Northington et al., "Ecosystem Function in Appalachian Headwater Streams during an Active Invasion by the Hemlock Woolly Adelgid," *PLOS One* 8, no. 4 (April 2013): 1.

21. Philip P. Kennedy, *The Blackwater Chronicle*, ed. Timothy Sweet (Morgantown: West Virginia University Press, 2002), 107–8.

22. Martin, "Ecosystem Dynamics in Central Appalachian Riparian Forests," 5–6.

23. Nash, *Blue Ridge 20/20*, 31.

24. Northington et al., "Ecosystem Function in Appalachian Headwater Streams," 1–8.

25. Gordon R. Thurow, "A New Subspecies of *Plethodon welleri*, with Notes on Other Members of the Genus," *American Midland Naturalist* 55, no. 2 (April 1956): 351–52; Charles F. Walker, "*Plethodon welleri* at White Top Mountain, Virginia," *Copeia* 1934, no. 4 (December 1934): 190.

26. Joe Mitchell and Whit Gibbons, *Salamanders of the Southeast* (Athens: University of Georgia Press, 2010), 264–66; Organ, "Studies on the Life History," 294.

27. Mitchell and Gibbons, *Salamanders*, 234–36.

28. Jenkins, "Impact of the Balsam Woolly Adelgid," 116; Rachel H. McManamay, Lynn M. Resler, James B. Campbell, Ryan A. McManamay, "Assessing the Impacts of Balsam Woolly Adelgid (*Adelges piceae* Ratz.) and Anthropogenic Disturbance on the Stand Structure and Mortality of Fraser Fir [*Abies fraseri* (Pursh) Poir.] in the Black Mountains, North Carolina," *Castanea* 76, no. 1 (March 2011): 1–19.

29. Martin, "Ecosystem Dynamics in Central Appalachian Riparian Forests," 28–29; Northington et al., "Ecosystem Function in Appalachian Headwater Streams," 1–8.

INDEX

Gilpatrick, Dorothy, 165
ginseng, 48, 121
gold: decline of Appalachian mining, 71–72; discovery in Georgia, 55–56; economic importance of, 54, 61, 64; environmental consequences of mining, 53, 62–63, 69–70, 72; expenses of mining, 60–61; lottery auctions, 56–57; mining in California, 53, 64–66; mining in Georgia, 53, 57–61, 69–72, 72–73; mining in North Carolina, 56, 61, 70; slavery in mining, 60, 62
golf, 128
Grandfather Hotel, 118, 119, 121
Grandfather Mountain, N.C., 200, 210; agriculture on, 121, 123; biodiversity of, 201, 206–7; botanical collecting on, 33, 40, 117; endangered species on, 201–2, 207–8; invasive species on, 203, 204–6, 207, 208; logging on, 123–24, 129–31; potential national park status, 132–33; scenic attraction, 117, 133–35, 136
Grant, Ulysses S., 87
Gray, Asa, 38, 42–43, 46, 49
Great Migration, 110
great rhododendron, 208–9
Great Smoky Mountains, 2, 132, 161
Great Smoky Mountains National Park, 132, 162, 204
Groves, Leslie, 169

Harney, Will Wallace, 48, 127
Harriman, 101
Hartley, John, 127
health resorts, 115, 116–17, 119–21, 127
hellbender. *See under* salamanders
hemlock, 204–7, 208–9, 210
hemlock woolly adelgid (HWA), 204–7, 209, 210
Hendrix, John, 158, 169, 176–77
High Country, 118, 135, 136
higher education, 176
Highlands, N.C., 115, 120, 124
Hiroshima, 158, 159
Hodge, James T., 68
Holston Ordnance Works, 165
Hotchkiss, Jedediah, 97, 98, 104, 105
Houston, T. J., 109–10
Huntington, Collis P., 103
hydraulic fracturing. *See* fracking

hydraulic mining, 54–55, 65–67, 69–72, 229n89. *See also* gold

iLoveMountains.org, 196
imidacloprid, 205, 209
Indian Removal Act, 56
invasive species, 200, 202–6, 207–9
Irwin, Bonita, 164–65

Jackson, Charles, 68
Jardin des Plantes (Paris), 37, 42
Johnson, Lyndon Baines, 193–94
Johnson, Moses, 109–10
Jones, John B., 74, 87, 90
Jones, Samuel, 87

K-25, 165, 166, 170, 171, 173
Kelsey, Harlan, 131–33
Kelsey, Samuel T., 115, 120, 124, 131, 243n79
Kennedy, Philip, 206
Kephart, Horace, 49–50, 132
Kilgore, Frank, 186
Killebrew, J. B., 143–44, 150
Kimball, Frederick, 105
King, Edward, 91, 97–98
King, William, 77, 80
Kingston Fossil Plant, 196
Knott County, Ky., 186
Knoxville, Tenn., 140, 169

Last Mountain, The, 196
Lavender, William, 111
Lenoir, Walter W., 124
Linville, N.C.: advertising of, 125–26, 128, 134; architecture, 125; founding, 115, 124–25; healthy nature of, 117, 128; recreation at, 125, 127–28, 133; transportation to, 128–29
Linville Improvement Company, 125–26, 128, 129, 133, 134
Little Switzerland, N.C., 135
livestock, 48, 121, 140
logging, 121–22, 129–31, 135
lynching, 111
Lyon, John, 33, 37–38, 39–40, 44
Lyon, Margene, 169

MacRae, Donald, 124
MacRae, Hugh, 124–25, 132